The Conversation on Biotechnology

Critical Conversations

Martin LaMonica, Series Editor

The Conversation U.S. is an independent, nonprofit news organization dedicated to delivering expert knowledge to the public through journalism. Every day The Conversation produces 10–12 stories through a collaboration between scholars and editors, with the scholars writing explanatory journalism and analysis based on their research and the editors helping them translate it into plain language. The articles can be read on TheConversation.com and have been republished by more than a thousand newspapers and websites through a Creative Commons license, meaning that the content is always free to read and republish.

The book series Critical Conversations is published collaboratively by The Conversation U.S. and Johns Hopkins University Press. Each volume in the series features a curated selection of subject-specific articles from The Conversation and is guest-edited by an expert scholar of the subject.

■

The Conversation on Biotechnology is guest-edited by Marc Zimmer, the Jean C. Tempel '65 Professor of Chemistry at Connecticut College and the author of *The State of Science: What the Future Holds and the Scientists Making It Happen* (Prometheus, 2020), *Illuminating Diseases: An Introduction to Green Fluorescent Proteins* (Oxford University Press, 2015), and four books for young adults.

Martin LaMonica is Executive Editor and Project Manager.

Kira Barrett is Editorial Assistant.

Gita Zimmerman is Illustrator of the part-title images, overseen by Conversation Marketing and Communications Manager Anissa Cooke-Batista.

Beth Daley is Editor and General Manager of The Conversation U.S.

Bruce Wilson is Chief Innovation and Development Officer of The Conversation U.S.

We would like to express our gratitude to the editors and scholars who produced the articles collected here and to thank our colleagues and funders who allow us to do this important work in the public interest.

THE CONVERSATION
on Biotechnology

edited by Marc Zimmer

Johns Hopkins University Press

BALTIMORE

Johns Hopkins University Press
2715 North Charles Street
Baltimore, Maryland 21218
www.press.jhu.edu

Library of Congress Cataloging-in-Publication Data

Names: Zimmer, Marc, editor.
Title: The conversation on biotechnology / Marc Zimmer.
Description: Baltimore, Maryland : Johns Hopkins University Press, 2023. |
    Series: Critical conversations | Includes bibliographical references and index.
Identifiers: LCCN 2022030338 | ISBN 9781421446141 (paperback ; acid-free
    paper) | ISBN 9781421446158 (ebook)
Subjects: LCSH: Biotechnology. | Medical technology. | Genetic engineering. |
    Food—Biotechnology. | BISAC: SCIENCE / Biotechnology | MEDICAL /
    Biotechnology
Classification: LCC TP248.24 .C68 2023 | DDC 660.6—dc23/eng/20221006
LC record available at https://lccn.loc.gov/2022030338

A catalog record for this book is available from the British Library.

Special discounts are available for bulk purchases of this book. For more
information, please contact Special Sales at specialsales@jh.edu.

Contents

Part II.
Biotechnology, Food, and the Environment   58

Part III.
Powerful Tools for Medicine and Health   114

Part IV.
Genetic Frontiers and Ethics   170

First, a Quick Story

IN THE FALL OF 2014, my wife sent me a listing for an environment and energy editor job at a place called The Conversation. I had vaguely heard of this outfit, one of many startups trying to innovate within the troubled world of journalism. As a longtime reporter and editor, I had little optimism that anyone could fix the media's broken business model. But I was intrigued by the organization's approach of having college professors and researchers fill the gap left by layoff after layoff of experienced reporters. I emailed the managing editor and landed an interview.

Days after I left the interview held in a cramped base-ment office on the Boston University campus, the mission

underpinning the venture—to improve the public discourse—
stayed with me. Could academics, working with journalists,
help fulfill the vital role of journalism to inform the public by
sharing more facts and knowledge?

As I now write this foreword years later, I can say that the
power of The Conversation's founding idea and novel editorial
model endures. And millions of people benefit every day. A
media nonprofit with editions in multiple countries, we publish
daily news analysis and explanatory journalism that comes out
of a collaboration between academics and journalists. Put
another way, it's like a digital newspaper whose reporters are
researchers and professors with deep subject expertise and
whose editors are journalists who themselves are topic
specialists with years of experience covering the news.

We have a website, multiple email newsletters, and many
outposts on social media. Why make a book from our daily
journalism? And who would want to read it?

As an independent media nonprofit, we exist to serve the
public with accurate and reliable information you can use to
navigate an increasingly complex world. We are funded by
universities, foundations, and individual donors. This collection's contributions are grounded in academic research, since
the authors—sometimes people who have worked in their field
for decades—are writing about their areas of expertise. There
are citations to peer-reviewed articles and books throughout,
and this collection itself has gone through peer review.

But this rich information is also accessible. Working with
journalist-editors, the academics write with a general reader
in mind—anyone who is curious to learn more about a subject
and values the knowledge that comes from years of study and
academic achievement. The format is what you'd expect to

read in most media outlets as well: stories of somewhere between 800 and 1,200 words long that can be read in a brief amount of time. Since many of the chapters in this collection were originally written in response to events in the news, most have been modified slightly to remove references that would date them. Inevitably, new developments or events will supersede what we published previously, but we have tried to ensure the chapters are accurate.

The themes we have chosen for these collections reflect our editorial values and the makeup of our newsroom. At The Conversation, we have the luxury of having dedicated editors in a wide range of subjects, from education to climate change—something that few media outlets still have the resources to do.

Perhaps the biggest payoff from this multidisciplinary approach comes when we examine one subject in depth, as we do in this book. It's enlightening, and often quite fun, to discover what you didn't know by reading an essay from an ethicist, for instance, a few pages after hearing what a microbiologist or a historian has to say. Having multiple entry points into a subject helps us connect the dots to see the bigger picture. Also, solving thorny societal problems requires a wide-angle view: discussions about new technologies and scientific developments, for example, cannot be separated from the societal impacts they have, just as, say, dealing with environmental challenges requires input from the physical sciences, politics, sociology, and more.

It's worth noting that the multiple voices and different viewpoints in this collection are just that—varied. That means the tone of chapters will vary, and some authors may not even agree on certain points. But that's OK, because the goal is to give you, the reader, the context and a foundational under–standing of issues that are important to living in today's society.

Our hope is that, after reading this collection, you will feel better equipped to make sense of news in the headlines or to grasp the significance of new research. In other cases, you may just be entertained by a good story.

Martin LaMonica

The Jigsaw Puzzle and the "Gene Gap"

The Process of Doing Science

The easiest and most understandable way of presenting science is often not the correct way. I love telling stories. My research seminars follow a distinct path with a beginning, middle, and end. As we follow the research, there are many opportunities to stop along the way and appreciate the scenic vistas (the work of others) lining the research path and to admire the path taken. There is always a defined destination: the goal of the seminar is to take my audience to the equivalent of a spectacular waterfall; in my case it is often something like "and that is why this particular fluorescent protein is so bright." This is the way most scientists and science journalists present scientific research—as a journey with a distinct end point—no matter whether the presentation is a peer-reviewed paper, seminar lecture, or newspaper article.

That is not, however, how research is done. In reality, experimental results are scattered about the research landscape, and scientists struggle to connect the experimental evidence from a starting point to the desired destination with a single narrative. Most paths lead nowhere: they are dead ends. If the evidence doesn't exist, then new experiments have to be undertaken. Another common misconception is that science is like simple mathematics with proofs and irrefutable facts, such as $5 + 3 = 8$. When science does not follow a direct path to discovery or conform with the invariability of simple mathematics, the general public may lose faith in the process.

Assembling a jigsaw puzzle may be a much more accurate analogy for scientific research than the linear story line typically used. Nature is a cruel puzzle maker. She presents us with massive and complex puzzles but withholds the box, leaving researchers with no idea of the big picture or the number of puzzle pieces. When nature presents a new puzzle, such as COVID-19, interested researchers face a blank slate—all they have to go on is information obtained by examining older related puzzles; in the case of COVID-19, this information came from previous outbreaks of SARS (severe acute respiratory syndrome) and MERS (Middle East respiratory syndrome). Researchers do experiments, and when they think they have enough information, they publish the results in a peer-reviewed journal. They are telling the world that they have a puzzle piece. Important puzzle pieces that join with existing pieces or reveal important information about the puzzle are published in the most prestigious journals (*Nature, Science*, and *Cell*).

With any new puzzle there is a significant amount of uncertainty about where pieces belong and about their relationship to other pieces. Scientists tend to be cautious about extrapolating from their data. Just because a scientist has a puzzle piece clearly showing part of an anchor chain, say, doesn't mean that chain is attached to a yacht, which in turn is moored in a Mediterranean harbor. The anchor chain piece could actually be a paperweight, and the puzzle is not a Mediterranean scene at all. Call it a case of true piece, false big-picture.

That is why we have seen so much uncertainty and change in our understanding of COVID-19. The science has not been incorrect; it is just that the context is unknown. The COVID-19 pandemic has placed the scientific process under intense scrutiny, and because of misconceptions of how science is done, some people have decided that science has not met their expectations. This void of certain knowledge has tempted some scientists and many nonscientists to force incorrect pieces into the COVID-19 puzzle. This can be disastrous when the puzzle pieces are fabricated or misplaced, whether by accident or on purpose; they slow down and misdirect research and misinform the public (for example, by positing that hydroxychloroquine cures COVID-19 or that vaccines cause autism). The increased complexity and magnitude of scientific knowledge, as well as the plethora of informational sources on the internet, make it challenging to differentiate scientifically established facts from misinformation when assembling new puzzles.

In older puzzles, by contrast, the picture may be nearly complete, so adding a piece normally involves little uncertainty.

The position of the puzzle piece is not in doubt, as it is flanked by locked-in pieces and its contribution to the big picture was foreseen. Climate change is one such puzzle. It is coming together; we understand what is happening and have a great overview of the big picture. For scientists, there is no more uncertainty about the picture under construction.

In the world of biotechnology, there are puzzle pieces being released every day and new puzzles being started, reflecting the rapid change we're living through. What is the big picture starting to look like? Where might biotech be heading?

The Growth of Science and Biotechnology

Science is growing faster and faster, and most of the growth is due to the development of new techniques. With these new techniques we can not only generate more data (puzzle pieces) than ever before, but we are also using machine learning to process all the new information, resulting in a more nuanced understanding of the complexities of science, which ultimately births new applications. Many of these techniques are applicable to biotechnology. Its growth is driven by the maturation of many new methods such as CRISPR, imaging, microscopy, optogenetics, and artificial intelligence.

In 2007 researchers employed by Danisco, a Danish food and beverage company, were the first to prove that CRISPR is a bacterial defense system. They were interested in CRISPR because *Streptococcus thermophiles*, a milk-fermenting bacterium used by Danisco and other dairy processors to make yogurt and mozzarella and other cheeses, is vulnerable to viral attack. According to the US Department of Agriculture, 1.02 billion kilograms of mozzarella cheese and 621 million kilograms of yogurt are produced annually with *S. thermophiles*. Viral infections of *S. thermophiles* are the largest cause of incomplete fermentation and lost production in the dairy industry. Most dairy products contain some *S. thermophiles*, and humans ingest over a billion trillion live *S. thermophiles* a year.[1] Today many manufacturers use *S. thermophiles* cultures that have been genetically modified with CRISPR sequences to protect them from the most common viral outbreaks. According to Rodolphe Barrangou, who did this research at Danisco USA, "If you've eaten yogurt or cheese, chances are you've eaten CRISPRized cells."[2] Yet most of us have not heard of this, the first commercial use of CRISPR. It is an instance where CRISPR is coopted as a defense system and is not acting as a gene editor.

This example is typical in that the bulk of biotechnological advances are invisible to us. We are acquainted only with those areas of immediate interest to us or to the spectacular advances picked up by traditional and social media, such as vaccines using mRNA (messenger RNA), pig heart transplants into humans, sickle cell anemia treatments, and the Theranos debacle.

Science and Biotech Are Incredibly Powerful

Science is not only advancing more quickly, but it is increasingly more potent as well. CRISPR, machine learning, optogenetics, and gene drives are powerful techniques that will lead to, indeed have already led to, biotechnologies that will change our lives and the lives of our children. Quite justifiably, CRISPR and artificial intelligence (AI) have received the most attention. As a computational chemist interested in the structural properties of biomolecules, I have been closely following advances in both AI and CRISPR. AI has changed the way science is done, as it allows scientists to analyze the massive amounts of data that modern instrumentation generates. It can find a needle in a million haystacks of data, and in a process called machine learning, AI can learn from the data it analyzes. AI is accelerating our advances in gene hunting, drug design, and organic synthesis. It has solved one of the grand challenges in science, the protein folding problem.

Thanks to programs like AlphaFold and RoseTTAFold, researchers like me can obtain the three-dimensional structure of proteins from the sequence of amino acids that make up the protein—for free and in an hour or two. Before AlphaFold we had to crystalize the proteins and solve the structures by X-ray crystallography, a process that took months and cost tens of thousands of dollars per structure. Many small biotech companies have been founded to use AI to design new, small pharmaceutical molecules that selectively bind particular enzymes, and the big drug companies are all developing their own versions of AlphaFold. Perhaps this will be the big step toward targeted drug design that we have all been looking for.

In July 2018, on *Last Week Tonight with John Oliver*, John Oliver turned his wit to CRISPR, saying, "One of the most extraordinary aspects of CRISPR-directed gene editing is the broad impact it has had on so many disciplines. It seems at this point that the only thing CRISPR won't be able to do is open your refrigerator door!"[3] CRISPR is a bacterial defense system that has been modified to be a general-purpose molecular word processor. Jennifer Doudna and Emmanuelle Charpentier were awarded the 2020 Nobel Prize in Chemistry for developing CRISPR as a method for editing the genome. It allows us to find not just a gene but a specific part of the gene, then change it, delete it, make the cell express more or less of the gene's product, or even substitute a foreign gene. It is Microsoft Word for the book of life.

Genetic modifications that used to take sophisticated biological laboratories years to do can now be done in days. A complicated research project costing hundreds of thousands of dollars has become an undergraduate project costing hundreds of dollars. CRISPR has revolutionized biotech. Before CRISPR, gene editing was species-specific, expensive, and tedious. Now researchers can edit genes in any species, and it's cheap and fast.[4] This book has a number of chapters on CRISPR-associated applications. This reflects CRISPR's importance to biotechnology.

The domestication of the first agricultural plants can be traced back at least 11,000 years. Since then we have been using traditional plant breeding techniques that rely largely on selective and crossbreeding methods. On March 18, 2018, the US Department of Agriculture issued a statement saying that the agency would not regulate genetically edited plant varieties that were indistinguishable from plants that could be obtained by traditional breeding methods. Since *unregulated* means *unlabeled*, and since CRISPR-modified plants will be produced much more quickly and less expensively than those created by traditional breeding methods, I am sure that in five years' time I will be eating cheaper and healthier CRISPR'd rice, potatoes, and tomatoes. They will be indistinguishable from conventional foods, and because their gene edits are not regulated or labeled, I won't know the difference. Part II of this book focuses on biotechnology in the food and agricultural sectors, an underreported area of biotech. The amount one hears and reads about biotech in the agricultural sector is disproportionally low compared with its relevance and importance to the public, probably because it benefits agribusiness to stay under the radar and because agriculture is not as sexy and interesting to the average reader as medicine.

There are more than 10,000 genetic disorders caused by mutations that occur on only one gene: these are the so-called single-gene disorders. They affect millions of people. Sickle cell anemia, cystic fibrosis, and Huntington's disease are among the most well known of these disorders. They are all obvious targets for CRISPR therapy, and numerous human trials are currently under way to cure single-gene disorders such as Leber congenital amaurosis, a severe eye disease causing visual impairment. Despite all the good news, we must temper our expectations for CRISPR in the medical arena, as most diseases are not caused by single-gene defects.

The Power of Biotech Is Associated with Risk

In 1926 J. B. S. Haldane wrote an essay titled "On Being the Right Size." Drop a mouse from a building, he writes, and it survives, whereas "a rat is killed, a man is broken, a horse splashes" (p. 424).[5] Gravity limits our size; it is the enemy of the large. Elephants have massive hearts to pump oxygen-bearing blood through their bodies, and they have thick, sturdy bones to remain upright. The larger the organism, the greater its complexity—just compare a fly with an eagle. But complexity only gets you so far. To get larger you have to change the conditions, such as going from a terrestrial to a marine environment. I suspect there is a right size for everything, including science. The big question is, What is the right size for science?

One of my favorite books is Ian Goldin's cleverly titled book *The Butterfly Defect: How Globalization Creates Systemic Risks, and What to Do about It*, coauthored with Mike Mariathasan.[6] The book was published in 2014, and it eerily forewarned of a global pandemic, caused by a spillover disease from wet markets in Chinese megacities, which would hit New York City and London particularly hard. It also warned of the power of the internet and social media to multiply the complexities of managing a pandemic, and it predicted shortages in the global supply chain, all of which have come true, of course, with the COVID-19 pandemic. Goldin is an economist and was an advisor to Nelson Mandela. I am an avid fan of his thinking. In the book the coauthors warn that "the corollary of progress is risk" (p. 204).

CRISPR, genetic engineering, and machine learning are massively powerful technologies that can be harnessed to benefit society, but their advantages are often offset by associated risks. Their advances need to be limited when they are dangerous and when they cross a commonly accepted ethical boundary. Currently, when an ethical dilemma arises from biotechnology—such as the ability to genetically modify human embryos—there is no standard review procedure in place to deal with it. Instead, ad hoc solutions are found. Is this sustainable?

We have to expect that once we have allowed gene editing for certain genetic disorders, the techniques and knowledge gained from this work will be used to genetically enhance embryos. In *A Crack in Creation*, Jennifer A. Doudna and Samuel H. Sternberg list some single-gene mutations that might tempt future parents: the *EPOR* gene confers increased endurance, the *LRP5* gene is responsible for extra strong bones, *DEC2* lowers the sleep requirements of people, and *MSTN* controls muscle growth.[7] As the possibility of CRISPR-designed babies comes closer to realization, the search to find more simple enhancements like these will accelerate. And genetics, like all of science, is growing faster and faster. It took seven years to sequence the first 1 percent of the human genome, then just another seven more to sequence the remaining 99 percent. By the time you read this book, the list of single-gene improvements will have grown. CRISPR-aided gene editing will widen existing wealth inequities, resulting in genetic inequities and creating what Doudna and Sternberg call the "gene gap." This is the first time in human history that not only will the rich have better lives, but their offspring will also have the opportunity to be better people, with stronger bones, more endurance, and reduced sleep requirements.

A *gene drive* forces a genetic trait through a population, defying the usual rules of inheritance. A gene drive is a form of genetic engineering that can radically raise the probability that genes will be inherited by the next generation. Normally there is roughly a 50 percent chance that a genetic trait will be passed from a given parent to that parent's offspring; with a gene drive, however, that possibility can approach near certainty. With the combination of CRISPR and gene drives, we have overcome constraints imposed by nature. We are no longer limited by the rules of inheritance. Evolution is in our hands; by using CRISPR, we can control a genetically defined trait and push it through a population with a gene drive. This is an incredibly powerful technique that has the potential to save millions of lives by, for example, eradicating malaria–carrying mosquitoes.

Yet we will soon reach the point where, for the first time in history, one application of genetic technology could potentially wipe out a whole species. How do we regulate this danger? How are we going to curb the weaponization of CRISPR gene drives when CRISPR and gene drives are used in labs all over the world and when the production of super–charged viruses and bacteria will be relatively cheap and will require minimal infrastructure? Perhaps this is a case where we will need to limit both the practice of science and the distribution of scientific knowledge.

In my opinion, science in itself is healthy. Nature's secrets are steadily being unpacked, our understanding of the universe is expanding, and we are continuing to build our knowledge. But, and this is a big but, I think the future of science is too bright just to wear shades.[8] Our scientific abilities and knowledge are increasing at a faster and faster rate; as a consequence, science is outgrowing its supporting structures. Governments, funding agencies, philosophers, and scientists themselves are struggling to find ways to make the most of CRISPR, of stem cells, and of machine learning, while keeping all the potential abuses at bay. Science and technology have enabled robust expansion for humans. We have overcome constraints imposed by nature and evolution. We now have no predators—apart from ourselves—to limit our growth. We can impose our will on nature and are in charge of our own destiny. One only has to look at climate change, however, to realize that we are not doing a good job of ensuring the future for our children.

To be responsible citizens, we all need to be cognizant of what is happening in science and biotech and to remember the words of Goldin and Mariathasan: "the corollary of progress is risk." We need to read books, like this one, that present the thoughts of experts who do science, books in which opinions were invited to address what is happening in today's biotechnology from a variety of perspectives and where the aim is not to achieve a false balance in which equal space is given to discredited fringe ideas alongside substantiated scientific ones.

Notes

1. Doudna, J. A., & Sternberg, S. H. (2017). *A crack in creation: Gene editing and the unthinkable power to control evolution.* Houghton Mifflin Harcourt.
2. Zimmer, C. (2015, April 6). Breakthrough DNA editor born of bacteria. *Quanta Magazine.* https://www.quantamagazine.org/crispr-natural-history-in-bacteria-20150206/.
3. Season 5, Episode 17. (2018, July 1). Gene editing [segment of episode]. In *Last night with John Oliver*; HBO. https://www.hbo.com/video/last-week-tonight-with-john-oliver/seasons/season-5/episodes/77-episode-136/videos/july-1-2018-gene-editing.
4. Zimmer, M. (2020). *The state of science: What the future holds and the scientists making it happen.* Prometheus.
5. Haldane, J. B. S. (1925, December 1). On being the right size. *Harper's Monthly Magazine, 152.*
6. Goldin, I., & Mariathasan, M. (2014). *The butterfly defect: How globalization creates systemic risks, and what to do about it.* Princeton University Press.
7. Doudna & Sternberg, *A crack in creation.*
8. I am alluding here to a one-hit wonder of the 1980s. Timbuk 3. (1986). The future's so bright, I gotta wear shades." On *Greetings from Timbuk3.* I.R.S.

The Conversation on Biotechnology

Part I.

Building Blocks of Life

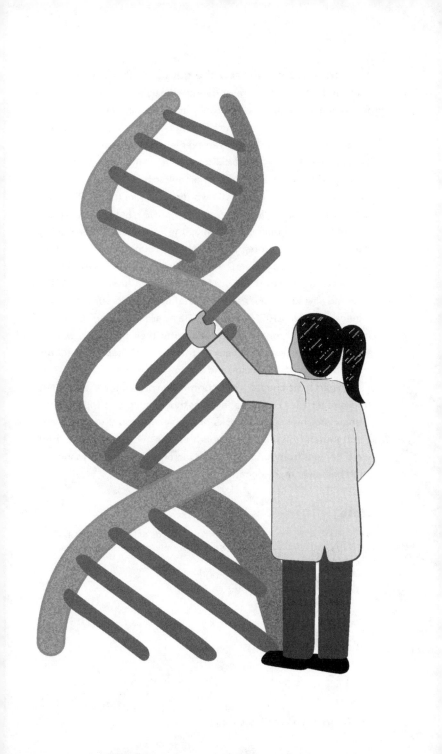

Humans have influenced the genes of other organisms for thousands of years by domesticating animals and cultivating plants for food and other uses. But today scientists and engineers have enormous power to wield the building blocks of life—genes, proteins, and other molecules—in an ever-expanding array of applications. Some researchers are calling the 21st century the "century of biology," driven by transformational tools in bioscience.

These tools act as genetic word processors (CRISPR), facilitate new ways of delivering vaccines (mRNA), activate individual neurons (optogenetics), and light up the internal workings of living cells (fluorescent proteins). The Nobel Foundation referred to CRISPR as "genetic scissors: a tool for rewriting the code of life" in its announcement of the 2020 Nobel Prize in Chemistry.[1] CRISPR is already living up to its early reputation, but are our long-term expectations for CRISPR realistic? Be aware that not all transformational breakthroughs have lived up to their hype; it is worthwhile, for instance, to look back on the Human Genome Project for a fact check on its promise—and accomplishments.

In part I, the chapters offer primers on proteins, genes, mRNA (messenger RNA), cloning, and methods, such as CRISPR, that manipulate the building blocks of life. A common thread running through most of the chapters is that the techniques described were derived from nature through incremental work by many researchers over many decades.

Note

1. *Genetic scissors: A tool for rewriting the code of life*. (2020). Nobel Prize. https://www.nobelprize.org/prizes/chemistry/2020/popular -information/.

Chapter 1

What Is mRNA? The Messenger Molecule That's Been in Every Living Cell for Billions of Years Is of Great Interest to Vaccine Developers

PENNY RIGGS

ONE SURPRISING STAR of the coronavirus pandemic response has been the molecule called mRNA, short for messenger RNA. It's the key ingredient in the Pfizer–BioNTech and Moderna COVID–19 vaccines. But mRNA itself is not a new invention from the lab. It evolved billions of years ago and is naturally found in every cell in your body. Scientists think RNA originated in the earliest life forms, even before DNA existed.[1]

Here's a crash course on just what mRNA is and the important job it does.

Meet the Genetic Middleman

You probably know about DNA. It's the molecule that contains all of your genes spelled out in a four-letter code: A, C, G, and T. DNA is found inside the cells of every living thing. It's protected in a part of the cell called the nucleus. The genes are the details in the DNA blueprint for all the physical characteristics that make you uniquely you.

But the information from your genes has to get from the DNA in the nucleus out to the main part of the cell—the cytoplasm—where proteins are assembled. Cells rely on proteins to carry out the many processes necessary for the body to function. That's where mRNA comes in.

Sections of the DNA code are transcribed into shortened messages that are instructions for making proteins. These messages—the mRNA—are transported out to the main part of the cell. Once the mRNA arrives, the cell can build particular proteins from these instructions.

The structure of RNA is similar to DNA but has some important differences. RNA is a single strand of code letters (nucleotides), while DNA is double-stranded. The RNA code contains a U instead of a T—uracil instead of thymine. Both RNA and DNA structures have a backbone made of sugar and phosphate molecules, but RNA's sugar is ribose and DNA's is deoxyribose. DNA's sugar contains one less oxygen atom, and this difference is reflected in their names: DNA is the nickname for deoxyribonucleic acid, and RNA stands for ribonucleic acid.

Identical copies of DNA reside in every single cell of an organism, from a lung cell to a muscle cell to a neuron. RNA is produced as needed in response to the dynamic cellular environment and the immediate needs of the body. It's mRNA's job to help fire up the cellular machinery to build the proteins, as encoded by the DNA, that are appropriate for that time and place. The process that converts DNA to mRNA to protein is the foundation for how a cell functions.

Programmed to Self-Destruct

As the intermediary messenger, mRNA is an important safety mechanism in cells. It prevents invaders from hijacking the cellular machinery to produce foreign proteins, because any RNA outside the cell is instantaneously targeted for destruction by enzymes called RNases. When these enzymes recognize the structure and the U in the RNA code, they erase the message, protecting the cell from false instructions. Also, mRNA gives the cell a way to control the rate of protein production by turning the blueprints on or off as needed. No cell wants to produce every protein described in your whole genome all at once.

Messenger RNA instructions are timed to self-destruct, like a disappearing text or snapchat message. Structural features of mRNA—the U in the code, its single-stranded form, ribose sugar, and its specific sequence—ensure that the mRNA has a short half-life.[2] These features combine to enable the message to be "read," translated into proteins, and then quickly destroyed—within minutes for certain proteins that need to be tightly controlled or up to a few hours for others. Once the instructions vanish, protein production stops until the protein factories receive a new message.

Harnessing mRNA for Vaccination

All of mRNA's characteristics made it of great interest to vaccine developers. The goal of a vaccine is to get your immune system to react to a harmless version or part of a germ so that when you encounter the real thing, you're ready to fight it off. Researchers found a way to introduce and protect an mRNA message with the code for a portion of the spike protein on the SARS-CoV-2 virus's surface.[3]

The vaccine provides just enough mRNA to make a sufficient amount of the spike protein for a person's immune system to generate antibodies that will protect them if they are later exposed to the virus. The mRNA in the vaccine is soon destroyed by the cell, just as any other mRNA would be. The mRNA cannot get into the cell nucleus, and it cannot affect a person's DNA.

The first publicly available mRNA vaccines were authorized for emergency use to provide protection against COVID-19 in December 2020 by the US Food and Drug Administration (FDA). The Pfizer-BioNTech vaccine (Comirnaty) achieved full FDA approval in August 2021. The vaccines were considered "brand new" in 2020, but the underlying technology had been developed decades before and improved incrementally over time.[4] With knowledge from past research, rapid volunteer trials, and data collected in real time (from more than 500,000 mRNA vaccine

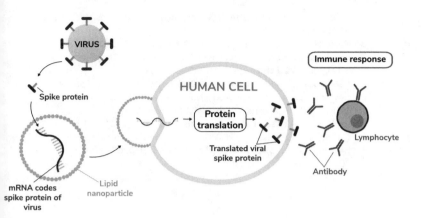

Spike protein

VIRUS

Immune response

HUMAN CELL

Protein translation

Translated viral spike protein

Lymphocyte

Antibody

mRNA codes spike protein of virus

Lipid nanoparticle

doses administered in the United States by the end of 2021), these vaccines were well tested for safety and effectiveness.

The success of these mRNA vaccines against COVID-19, in terms of both safety and efficacy, predicts a bright future for other vaccines that can be quickly tailored to meet new threats.[5] Early-stage clinical trials using mRNA vaccines have already been conducted for influenza, Zika, rabies, and cytomegalovirus. Certainly, creative scientists are already considering and developing therapies for other diseases and disorders that might benefit from an approach similar to that used for the COVID-19 vaccines.

Notes

1. Pearce, B. K., Pudritz, R. E., Semenov, D. A., & Henning, T. K. (2017). Origin of the RNA world: The fate of nucleobases in warm little ponds. *Proceedings of the National Academy of Sciences, 114*(43), 11327–11332. https://doi.org/10.1073/pnas.1710339114.
2. Sharova, L. V., Sharov, A. A., Nedorezov, T., Piao, Y., Shaik, N., & Ko, M. S. H. (2009). Database for mRNA half-life of 19,977 genes obtained by DNA microarray analysis of pluripotent and differentiating mouse

embryonic stem cells. *DNA Research, 16*(1), 45–58. https://doi.org/10.1093/dnares/dsn030.

3. Pardi, N., Hogan, M. J., Porter, F. W., & Weissman, D. (2018). MRNA vaccines—a new era in vaccinology. *Nature Reviews Drug Discovery, 17*(4), 261–279. https://doi.org/10.1038/nrd.2017.243.

4. Sahin, U., Karikó, K., & Türeci, Ö. (2014). MRNA-based therapeutics—developing a new class of drugs. *Nature Reviews Drug Discovery, 13*(10), 759–780. https://doi.org/10.1038/nrd4278.

5. Jackson, N. A., Kester, K. E., Casimiro, D., Gurunathan, S., & DeRosa, F. (2020). The promise of mRNA vaccines: A biotech and industrial perspective. *Npj Vaccines, 5*(1). https://doi.org/10.1038/s41541-020-0159-8.

Chapter 2

What Is CRISPR, the Gene Editing Technology That Won the Chemistry Nobel Prize?

DIMITRI PERRIN

IN 2020 THE ROYAL SWEDISH ACADEMY of Sciences awarded the Nobel Prize in Chemistry to Emmanuelle Charpentier and Jennifer Doudna for their work on CRISPR, a method of genome editing. A genome is the full set of genetic "instructions" that determine how an organism will develop. Using CRISPR, researchers can cut up DNA in an organism's genome and edit its sequence.

CRISPR technology is a powerhouse for basic research and is also changing the world we live in. There are thousands of research papers published every year on its various applications. These include accelerating research into cancers, mental illness, potential animal–to–human organ transplantation, better food production,[1] the elimination of malaria-carrying mosquitoes,[2] and the protection of animals from disease.

Charpentier is the director at the Max Planck Institute for Infection Biology in Berlin, Germany, and Doudna is a professor at the University of California, Berkeley. Both played a crucial role in demonstrating how CRISPR could be used to target DNA sequences of interest.

Taking Advantage of Bacterial Immunity

CRISPR technology is adapted from a system that is naturally present in bacteria and other unicellular organisms known as archaea. This natural system gives bacteria a form of acquired immunity. It protects them from foreign genetic elements (such as invading viruses) and lets them "remember" these in case they reappear. Like most advances in modern science, the discovery of CRISPR and its emergence as a key genome editing method involved efforts by many researchers, over several decades.

In 1987 Japanese molecular biologist Yoshizumi Ishino and his colleagues were the first to notice, in *E. coli* bacteria, unusual clusters of repeated DNA sequences interrupted by short sequences.[3] Spanish molecular biologist Francisco Mojica and colleagues later showed that similar structures were present in other organisms and proposed to call them CRISPR, an acronym for clustered regularly interspaced short palindromic repeats. In 2005 Mojica and other groups reported that the short sequences (or "spacers") interrupting the repeats were derived from other DNA belonging to viruses.[4] Evolutionary biologists Kira Makarova, Eugene Koonin, and colleagues eventually proposed that CRISPR and the associated Cas9 genes were acting as the immune mechanism. This was experimentally confirmed in 2007 by Rodolphe Barrangou and colleagues.[5]

These are just some of the key advances in a large body of research that ultimately led to discovering the potential of CRISPR as a genome editing tool.

A Programmable System

Among other CRISPR-associated genes, Cas9 encodes a protein that "cuts" DNA. This is the active part of the defense against viruses, as it destroys the invading DNA. In 2012 Charpentier and Doudna showed that the spacers acted as markers for guiding where Cas9 would make a cut in the DNA. They also showed that an artificial Cas9 system could be programmed to target any DNA sequence in a lab setting. This was a groundbreaking discovery, which opened the door for CRISPR's wider applications in research.

In 2013, for the first time, groups led by American bio-chemist Feng Zhang and geneticist George Church reported genome editing in human cell cultures using CRISPR-Cas9. It has since been used in countless organisms from yeast to cows, plants, and corals. Today, CRISPR is the preferred tool for gene editing for thousands of researchers.

A Technical Revolution with Endless Applications

Humans have altered the genomes of species for thousands of years by selective breeding. Genetic engineering—the direct manipulation of DNA by humans—has existed only since the 1970s. CRISPR-based systems fundamentally changed this field, as they allow for genomes to be edited in living organisms cheaply, with ease and with extreme precision.

CRISPR is currently making a huge impact in health. There are clinical trials of its therapeutic use for blood disorders such as sickle cell disease or beta-thalassemia, for the treatment of the most common cause of inherited childhood blindness (Leber congenital amaurosis), and for cancer immunotherapy.

CRISPR also has great potential in food production. It can be used to improve crop quality, yield, disease resistance, and herbicide resistance. When used on livestock, it can lead to better disease resistance, increased animal welfare, and improved productive traits, such as producing more meat, milk, or high-quality wool.

With Great Power . . .

A number of challenges for CRISPR technology remain, however. Some are technical, such as the risk of off-target modifications (which happen when Cas9 cuts at unintended locations in the genome). Other problems are societal.

CRISPR was infamously used in one of the most controversial experiments of recent years, when Chinese biophysicist He Jiankui unsuccessfully attempted to use the technology to modify human embryos and make them resistant to HIV (human immunodeficiency virus). This led to the birth of twins Lulu and Nana.

We need a broad and inclusive discussion on the regulation of such technologies, especially given their vast applications and potential. To quote CRISPR researcher Fyodor

Urnov, the work of Charpentier and Doudna has "changed everything."

Notes

1. Lemmon, Z. H., Reem, N. T., Dalrymple, J., Soyk, S., Swartwood, K. E., Rodriguez-Leal, D., Van Eck, J., & Lippman, Z. B. (2018). Rapid improvement of domestication traits in an orphan crop by genome editing. *Nature Plants, 4*(10), 766–770. https://doi.org/10.1038/s41477-018-0259-x.
2. Kyrou, K., Hammond, A. M., Galizi, R., Kranjc, N., Burt, A., Beaghton, A. K., Nolan, T., & Crisanti, A. (2018). A CRISPR–Cas9 gene drive targeting *doublesex* causes complete population suppression in caged *Anopheles gambiae* mosquitoes. *Nature Biotechnology, 36*(11), 1062–1066. https://doi.org/10.1038/nbt.4245.
3. Ishino, Y., Shinagawa, H., Makino, K., Amemura, M., & Nakata, A. (1987). Nucleotide sequence of the *iap* gene, responsible for alkaline phosphatase isozyme conversion in *Escherichia coli*, and identification of the gene product. *Journal of Bacteriology, 169*(12), 5429–5433. https://doi.org/10.1128/jb.169.12.5429-5433.1987.
4. Mojica, F. J. M., Díez-Villaseñor, C., García-Martínez, J., & Soria, E. (2005). Intervening sequences of regularly spaced prokaryotic repeats derive from foreign genetic elements. *Journal of Molecular Evolution, 60*(2), 174–182. https://doi.org/10.1007/s00239-004-0046-3.
5. Barrangou, R., Fremaux, C., Deveau Hélène, Richards, M., Boyaval, P., Moineau, S., Romero, D. A., & Horvath, P. (2007). CRISPR provides acquired resistance against viruses in prokaryotes. *Science, 315*(5819), 1709–1712. https://doi.org/10.1126/science.1138140.

Chapter 3

What Is a Protein? A Biologist Explains

NATHAN AHLGREN

A PROTEIN IS A BASIC ORGANIC STRUCTURE found in all of life. It's a molecule. And the key thing about a protein is that it's made up of smaller components called amino acids. I like to think of them as a string of differently colored beads. Each bead represents an amino acid, which is a smaller molecule containing carbon, nitrogen, oxygen, hydrogen, and sometimes sulfur atoms. So a protein is essentially a string composed of individual amino acids.

There are 22 different amino acids that can combine in any number of ways. A protein doesn't usually exist as a string, but instead it folds into a particular shape, depending on the order of the amino acids and how they interact. The resulting shape influences what the protein does in our body.

Where Do Amino Acids Come From?

The amino acids in our body come from the food we eat, and we also make some of them in our body. When we eat plant or animal proteins, our bodies take those large molecular chains and break them down into their individual amino acids. Once the proteins are broken down into amino acids in the digestive system, they are taken into our cells and kind of float around inside, as those beads in our analogy. Our cells can then remake them into any protein we need. We can make on our own about half of the amino acids we need, but we have to get the others from our food.

What Do Proteins Do in Our Body?

Scientists do not agree on their exact number, but most estimate that there are around 20,000 different proteins in our body. Some studies suggest, though, that there might be even more.[1] Proteins carry out a variety of functions, from doing metabolic conversions to holding cells together to causing muscles to work. Their functions fall into a few broad categories.

One functional category is structural. Your body is made up of many different kinds of structures; think of stringlike structures, globules, anchors, and the like. They form the stuff that holds your body together. For example, collagen is a protein that gives structure to your skin, bones, and even teeth.[2] Integrin is a protein that makes flexible linkages between your cells.[3] Your hair and nails are made of a protein called keratin.

Another big function for protein involves biochemistry—how your body carries out transformational reactions in its cells, like breaking down fat into simpler components or proteins into amino acids. This group of proteins are called enzymes. They take part in the chemical reactions necessary for life, making them happen more quickly and easily. They are generally involved in breaking down molecules or in forming new molecules from other smaller molecules. The enzyme pepsin helps break down proteins in the food we eat.[4] Likewise, there are proteins that are required to join amino acids together to make new proteins. Another example of a biochemical protein is hemoglobin, the protein that carries oxygen in the blood.[5]

Another category of proteins processes signals and information, like circadian clock proteins, which keep time in your cells.[6]

These are some of the main categories of functions that proteins carry out in our cells.

Why Is Protein Often Associated with Muscles and Meat?

Different types of foods have different kinds of protein content. In plants such as wheat and rice, there tend to be a lot of carbohydrates and less protein content. Meat in general has more protein content. A lot of protein is required to make the muscles in your body; proteins are also behind what makes your muscles move and work. That's why protein is often associated with eating meat and building muscle, but proteins are involved in much, much more than our muscles alone.

Notes

1. Poverennaya, E. V., Ilgisonis, E. V., Pyatnitskiy, M. A., Kopylov, A. T., Zgoda, V. G., Lisitsa, A. V., & Archakov, A. I. (2016). The size of the human proteome: The width and depth. *International Journal of Analytical Chemistry*, 1–6. https://doi.org/10.1155/2016/7436849.
2. *Molecule of the month: Collagen.* (n.d.). RCSB Protein Data Bank. https://pdb101.rcsb.org/motm/4.
3. *Molecule of the month: Integrin.* (n.d.). RCSB Protein Data Bank. https://pdb101.rcsb.org/motm/134.
4. *Molecule of the month: Pepsin.* (n.d.). RCSB Protein Data Bank. https://pdb101.rcsb.org/motm/12.
5. *Molecule of the month: Hemoglobin.* (n.d.). RCSB Protein Data Bank. https://pdb101.rcsb.org/motm/41.
6. *Molecule of the month: Circadian clock proteins.* (n.d.). RCSB Protein Data Bank. https://pdb101.rcsb.org/motm/97.

Chapter 4

Three Ways RNA Is Being Used in the Next Generation of Medical Treatment

OLIVER ROGOYSKI

YOU MIGHT HAVE HEARD ABOUT RNA RECENTLY, thanks to the development of the Pfizer and Moderna mRNA vaccines, which protect against COVID-19. But the potential medical uses of RNA molecules are much broader than vaccines.

RNAs, or ribonucleic acids, are some of the most important molecules for life on this planet. RNA is found in every cell in the body, and it plays an important role in the flow of genetic information. Messenger RNAs (mRNAs) copy and carry the genetic instructions from our DNA to the protein-making factories of our cells (ribosomes), which can then create the biological components and machinery that cells need. For example, actin proteins give cells their shape and structure and are crucial to muscle contraction. RNA also helps other biomolecules find one another and assists in bringing other proteins and RNAs together. These functions are crucial to managing the many levels of gene regulation, which is itself important for the proper functioning of the body.

RNA's wide range of capabilities, as well as its having a simple molecular sequence that can easily be read by researchers, has made it an extremely useful tool in the development of recent biomedical technologies—including CRISPR gene editing.[1] Here are three fields where RNA is being investigated.

Vaccines

The mRNA vaccines that are used to protect against SARS-CoV-2 (the virus that causes COVID-19) are the first of their kind to be licensed for widespread human use. But studies and clinical trials of RNA vaccines for other viruses—and even cancers—have been going on for a decade.[2] This type of vaccine introduces an RNA sequence into the body that causes its cells' ribosomes to express a specific, harmless viral protein temporarily, after which the foreign RNA molecules degrade. In turn, this trains the immune system to respond by producing strong protection against this virus should it encounter the virus again.

RNA vaccines are unlike conventional vaccines, which require either a harmless, inactive form of a virus or else small proteins or protein fragments made by a virus, to train the immune system. Designing and synthesizing an RNA sequence that provides the body with instructions is also easily and quickly done.

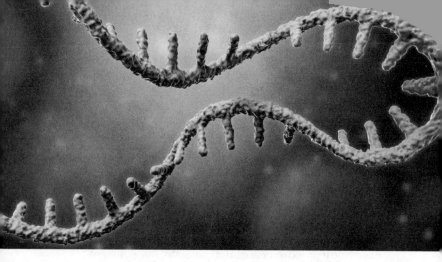

One of the biggest hurdles in making effective RNA-based drugs, though, has been the relative instability of the molecules. They degrade rapidly when exposed to certain common enzymes and chemicals and so must be kept at very low temperatures, such as the threshold of –70°C required for storing the Pfizer vaccine for COVID-19.[3]

Diagnostic Technologies

RNA is also playing an expanding role in diagnostics. Research into liquid biopsies (which only require a sample of human body fluids, such as blood) has increasingly shown that, by measuring levels of particular RNAs, many diseases can be diagnosed at an earlier stage, including cancers, neurodegenerative diseases, and cardiovascular disease.

Alongside making it easier and less invasive to collect samples, RNA biomarkers have additional advantages over tissue biopsies and other, more invasive, collection methods—such as skin, organ, or bone biopsies—because they're less painful and carry fewer risks. Combinations of RNA biomarkers can also be simultaneously evaluated, providing not only more confidence in a diagnosis but even allowing for predictions of disease progression and prognosis.[4] Large-scale studies that test the clinical suitability of this type of diagnostic tool are still needed, however.

Drug Development

RNA is also being used to help develop new drugs.

Drugs that target RNA can be identified, and in some instances customized, because researchers can sample RNA interactions and sequences linked to many different diseases from readily available databases. So far, drugs that target RNA have shown great promise in the treatment of rare diseases, such as Huntington's disease, which previously lacked effective treatments.

Drugs are also being designed that can target RNAs and modify or inhibit the function of certain genes or protein production, including those responsible for many diseases and symptoms. Several of these have been successfully used to treat viruses,[5] as well as neurodegenerative diseases, and have even been used in personalized medicine (treatments designed specifically for one patient).

RNA interference drugs are another area of research. These drugs silence a specific gene to treat a condition. Research into this type of drug is currently under way for many conditions, including amyloidosis (a rare disease caused by a buildup of proteins in the body), acute hepatic porphyria (a rare metabolic disorder), and several cancers (including lung cancer).

More recently, certain groups of RNAs and proteins have been shown to change the sensitivity of diseases (particularly cancers) to treatment. This has made some cancers less resistant to conventional treatment as a result.[6] This application could potentially provide a valuable new combination therapy for hard-to-treat diseases.

There has been plenty of investment in RNA therapeutics, and progress has been rapid over the last decade. With further clinical trials (testing safety and efficacy), and with improved methods for making these therapeutics at low cost and increasing their stability, we can soon hope to see even more results from this work: a whole new generation of medicines to use, which are more specialized and effective.

Notes

1. Jinek, M., Chylinski, K., Fonfara, I., Hauer, M., Doudna, J. A., & Charpentier, E. (2012). A programmable dual-RNA-guided DNA endonuclease in adaptive bacterial immunity. *Science, 337*(6096), 816–821. https://doi.org/10.1126/science.1225829.
2. Aarntzen, E. H., Schreibelt, G., Bol, K., Lesterhuis, W. J., Croockewit, A. J., de Wilt, J. H., van Rossum, M. M., Blokx, W. A., Jacobs, J. F., Duiveman-de Boer, T., Schuurhuis, D. H., Mus, R., Thielemans, K., de Vries, I. J., Figdor, C. G., Punt, C. J., & Adema, G. J. (2012). Vaccination with mRNA-electroporated dendritic cells induces robust tumor antigen–specific CD4+ and CD8+ T cells responses in stage III and IV melanoma patients. *Clinical Cancer Research, 18*(19), 5460–5470. https://doi.org/10.1158/1078-0432.ccr-11-3368.
3. Pardi, N., Hogan, M. J., Porter, F. W., & Weissman, D. (2018). mRNA vaccines—a new era in vaccinology. *Nature Reviews Drug Discovery, 17*(4), 261–279. https://doi.org/10.1038/nrd.2017.243.
4. Wen, G., Zhou, T., & Gu, W. (2020). The potential of using blood circular RNA as liquid biopsy biomarker for human diseases. *Protein*

& *Cell*, *12*(12), 911–946. https://doi.org/10.1007/s13238-020 -00799-3.

5. Chu, H., Kohane, D. S., & Langer, R. (2016). RNA therapeutics—the potential treatment for myocardial infarction. *Regenerative Therapy*, *4*, 83–91. https://doi.org/10.1016/j.reth.2016.03.002.

6. Iadevaia, V., Wouters, M. D., Kanitz, A., Matia-González, A. M., Laing, E. E., & Gerber, A. P. (2019). Tandem RNA isolation reveals functional rearrangement of RNA-binding proteins on *CDKN1B/p27*[Kip1] 3'UTRs in cisplatin treated cells. *RNA Biology*, *17*(1), 33–46. https://doi.org/10 .1080/15476286.2019.1662268.

Chapter 5

Why Sequencing the Human Genome Failed to Produce Big Breakthroughs in Disease

ARI BERKOWITZ

AN EMERGENCY ROOM PHYSICIAN, initially unable to diagnose a disoriented patient, finds on the patient a wallet-sized card providing access to his genome, or all his DNA. The physician quickly searches the genome, diagnoses the problem, and sends the patient off for a gene-therapy cure. That's what a Pulitzer Prize–winning journalist imagined that medicine in 2020 would look like when she reported on the Human Genome Project back in 1996.

A New Era in Medicine?

The Human Genome Project was an international scientific collaboration that successfully mapped, sequenced, and made publicly available the genetic content of human chromosomes—that is, human DNA. Taking place between 1990 and 2003, the project caused many to speculate about the future of medicine. In 1996, Walter Gilbert, a Nobel laureate, said, "The results of the Human Genome Project will produce a tremendous shift in the way we can do medicine and attack problems of human disease." In 2000, Francis Collins, then head of the Human Genome Project at the National Institutes of Health, predicted, "Perhaps in another 15 or 20 years, you will see a complete transformation in therapeutic medicine." The same year, President Bill Clinton stated that the Human Genome Project would "revolutionize the diagnosis, prevention and treatment of most, if not all, human diseases."[1]

It is now beyond 2020, yet no one carries a genome card. Physicians typically do not examine your DNA to diagnose or treat you. Why not? As I explain in an article in the *Journal of Neurogenetics*, the causes of common debilitating diseases are typically too complex to be amenable to simple genetic treatments, despite hope and hype to the contrary.[2]

Causation Is Complex

The idea that a single gene can cause common diseases has been around for several decades. In the late 1980s and early 1990s, high-profile scientific journals, including *Nature* and *JAMA*, announced single-gene causation of bipolar disorder, schizophrenia, and alcoholism, among other conditions and behaviors. These articles drew massive attention in the popular media but were soon retracted or had their findings fail attempts at replication.[3] These reevaluations completely undermined the initial conclusions, which often had relied on misguided statistical tests.[4] Biologists were generally aware of these developments, although the follow-up studies received little attention in the popular media.

There are indeed individual gene mutations that cause devastating disorders, such as Huntington's disease. But most common debilitating diseases are not caused by a mutation of a single gene. This is because people who have a debilitating genetic disease, on average, do not survive long enough to have numerous healthy children. In other words, there is a strong evolutionary pressure against propagating such mutations. Huntington's disease is an exception that endures because it typically does not produce symptoms until a patient is beyond their reproductive years. Although new mutations for many other disabling conditions occur by chance, they don't become frequent in the population.

Instead, most common debilitating diseases are caused by combinations of mutations in many genes, each having a very small effect. They interact with one another and with environmental factors, modifying the production of proteins from genes. The many kinds of microbes that live within the human body can play a role, too.

Since frequently occurring serious diseases are rarely caused by single-gene mutations, they cannot be cured by replacing the mutated gene with a normal copy, the premise for gene therapy. Gene therapy has gradually progressed in research along a bumpy path, which has included accidentally causing leukemia and at least one death,[5] but doctors recently have been successful treating some rare diseases in which a single-gene mutation has had a large effect.[6] Gene therapy for rare single-gene disorders is likely to succeed, but it must be tailored to each individual condition. The enormous cost and the relatively small number of patients who can be helped by such a treatment may create insurmountable financial barriers to its development. For many diseases, gene therapy may never be useful.

A New Era for Biologists

The Human Genome Project has had an enormous impact on almost every field of biological research by spurring technical advances that facilitate fast, precise, and relatively inexpensive sequencing and manipulation of DNA. But these advances in research methods have not led to dramatic improvements in the treatment of common debilitating diseases.

Even though you don't have a genome card to take to your next doctor's appointment, perhaps you can bring with you a more nuanced understanding of the relationship between genes and disease. A realistic understanding of disease causation may protect patients against beguiling stories and false promises.

Notes

1. Here are two papers that discuss unrealized aspirations for personalized medicine: Murray, J. (2012). Personalized medicine: Been there, done that, always needs work! *American Journal of Respiratory and Critical Care Medicine, 185*(12). https://doi.org/10.1164/rccm.201203 -0523ED; Joyner, M. J., & Paneth, N. (2019). Promises, promises, and precision medicine. *Journal of Clinical Investigation, 129*(3): 946–948. https://doi.org/10.1172/JCI126119.
2. Berkowitz, A. (2019). Playing the genome card. *Journal of Neurogenetics, 34*(1), 189–197. https://doi.org/10.1080/01677063.2019.1706093.

3. Cloninger, C. R. (1994). Turning point in the design of linkage studies of schizophrenia. *American Journal of Medical Genetics, 54*(2), 83–92. https://doi.org/10.1002/ajmg.1320540202.

4. Alper, J. S., & Natowicz, M. R. (1993). On establishing the genetic basis of mental disease. *Trends in Neurosciences, 16*(10), 387–389. https://doi.org/10.1016/0166-2236(93)90003-5.

5. Jenks, S. (2000). Gene therapy death—"everyone has to share in the guilt." *JNCI: Journal of the National Cancer Institute, 92*(2), 98–100. https://doi.org/10.1093/jnci/92.2.98.

6. Dunbar, C. E., High, K. A., Joung, J. K., Kohn, D. B., Ozawa, K., & Sadelain, M. (2018). Gene therapy comes of age. *Science, 359*(6372). https://doi.org/10.1126/science.aan4672.

Chapter 6

Editing Genes Shouldn't Be Too Scary—
Unless They Are the Ones That Get
Passed to Future Generations

ELEANOR FEINGOLD

GENE EDITING IS ONE OF THE SCARIER THINGS in science news, but not all gene editing is the same. It matters whether researchers edit "somatic" cells or "germ line" cells.

Germ line cells are the ones that propagate into an entire organism—either cells that make sperm and eggs (known as germ cells) or the cells in an early embryo that will later differentiate in their functions. What's critical about germ line cells is that a change or mutation in one will go on to affect every cell in the body of a baby that grows from them.

Somatic cells are everything else—cells in organs or tissues that perform a specific function. Skin cells, liver cells, eye cells, and heart cells are all somatic. Changes in somatic cells are much less significant than changes in germ line cells. If you get a mutation in a liver cell, you may end up with more mutant liver cells as the mutated cell divides and grows, but it will never affect your kidney or your brain. Our bodies accumulate mutations in somatic tissues throughout our lives. Most of the time we never know it or suffer any harm. The exception is when one of those somatic mutations grows out of control and leads to cancer.

I am a geneticist who studies the genetic and environmental causes of a number of different disorders, from birth defects, in the form of cleft lip and palate,[1] to diseases of old age like Alzheimer's.[2] Studying the genome always entails thinking about how the knowledge you generate may be used and about whether those likely uses would be ethical. So geneticists have been following the gene editing news with great interest and concern.

In gene editing, it matters enormously whether you are messing with a germ line cell, and thus the whole of an individual and any future descendants of that individual, or just one organ. Gene therapy—for fixing faulty genes in individual organs—has been one of the great hopes of medical science for decades.[3] There have been a few successes but many more failures. Gene editing may make gene therapy more effective, potentially curing debilitating or fatal diseases in

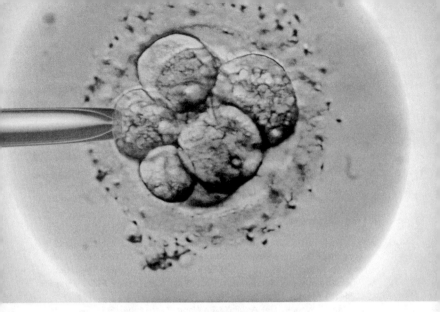

adults. The National Institutes of Health run a well-respected and highly ethical research program for developing tools for safe and effective gene editing to cure disease.[4]

But editing germ line cells that then create babies whose genes have been manipulated is a very different story, with multiple ethical issues. The first set of concerns is medical: at this point we don't know anything about the safety. "Fixing" the cells in the liver of someone who might otherwise die of liver disease is one thing, but "fixing" all of the cells in a baby who is otherwise healthy is a proposition of much higher risk. This is why the 2018 announcement that a Chinese scientist had done just that continues to generate vigorous debate among scientists and ethicists.[5]

But even if we knew the procedure was safe, gene editing of the germ line would still catapult us straight into all of the "designer baby" controversies and the problems of creating a world where people can micromanage their offspring's genes. It does not take much imagination to fear that gene editing could bring us a new era of eugenics and discrimination.

Does gene editing still sound scary? It should. But it makes a big difference whether you are manipulating individual organs or whole human beings.

Notes

1. Howe, L. J., Lee, M. K., Sharp, G. C., Davey Smith, G., St Pourcain, B., Shaffer, J. R., Ludwig, K. U., Mangold, E., Marazita, M. L., Feingold, E., Zhurov, A., Stergiakouli, E., Sandy, J., Richmond, S., Weinberg, S. M., Hemani, G., & Lewis, S. J. (2018). Investigating the shared genetics of non-syndromic cleft lip/palate and facial morphology. *PLOS Genetics, 14*(8). https://doi.org/10.1371/journal.pgen.1007501.
2. Yan, Q., Nho, K., Del-Aguila, J. L., Wang, X., Risacher, S. L., Fan, K.-H., Snitz, B. E., Aizenstein, H. J., Mathis, C. A., Lopez, O. L., Demirci, F. Y., Feingold, E., Klunk, W. E., Saykin, A. J., Cruchaga, C., & Kamboh, M. I. (2018). Genome-wide association study of brain amyloid deposition as measured by Pittsburgh Compound-B (PiB)-PET imaging. *Molecular Psychiatry, 26*(1), 309–321. https://doi.org/10.1038/s41380-018-0246-7.
3. Ginn, S. L., Amaya, A. K., Alexander, I. E., Edelstein, M., & Abedi, M. R. (2018). Gene therapy clinical trials worldwide to 2017: An update. *Journal of Gene Medicine, 20*(5). https://doi.org/10.1002/jgm.3015.
4. US Department of Health and Human Services. (2021, December 1). *Somatic cell genome editing.* National Institutes of Health. https://commonfund.nih.gov/editing.
5. Evans, J. H. (2021). Setting ethical limits on human gene editing after the fall of the somatic/germline barrier. *Proceedings of the National Academy of Sciences, 118*(22). https://doi.org/10.1073/pnas.2004837117.

Chapter 7

How Many Genes Does It Take to Make a Person?

SEAN NEE

WE HUMANS LIKE TO THINK of ourselves as atop the heap compared with all the other living things on our planet. Life has evolved over three billion years from simple one-celled creatures through to multicellular plants and animals that come in all shapes and sizes and abilities. In addition to growing ecological complexity, over the history of life we've also seen the evolution of intelligence, complex societies, and technological invention, until we arrive today at people cruising 35,000 feet above sea level and chatting about the in-flight movie.

It's natural to think of the history of life as progressing from the simple to the complex and to expect this to be reflected in increasing gene numbers. We fancy ourselves as leading the way with our superior intellect and global domination, so our expectation may be that, since we're the most complex creature, we must also have the most elaborate set of genes.

This presumption seems reasonable, but the more researchers find out about other genomes, the more flawed it appears to be. About half a century ago the estimated number of human genes was in the millions. Today we're down to about 20,000.[1] We now know, by way of comparison, that bananas, with their 30,000 genes, have 50 percent more genes than we do.[2]

As researchers devise new ways to count not just the genes an organism has, but also the ones it has that are supernumerary, there's a clear convergence between the number of genes in what we've always thought of as the simplest life-forms—viruses—and the most complex: us. So it's time to rethink the notion that the complexity of an organism is reflected in the size of its genome.

Counting Up the Genes

We can think of all our genes together as the recipes in a cookbook for making us. They're written with the bases of DNA, abbreviated with the letters A, C, G, and T. The genes provide instructions for how and when to assemble the proteins that we're made of and that carry out all the functions of life in our bodies. A typical gene requires about 1,000 letters. Together with the environment and experience, genes are responsible for what and who we are, so it's interesting to investigate how many genes add up to a whole organism.

When we're talking about numbers of genes, we can tally the actual count for viruses but only estimates for human beings, for an important reason. One challenge for counting genes in eukaryotes—which include us, bananas, and yeast like *Candida*—is that our genes are not lined up like ducks in a row. Our genetic recipes are arranged as though the cookbook's pages have all been ripped out and mixed up with three billion other letters, about 50 percent of which actually describe inactivated, dead viruses. So, in eukaryotes, it's hard to count up the genes that have true functions and separate them from what's extraneous.

In contrast, counting genes in viruses—and bacteria, which can have 10,000 genes—is relatively easy.[3] This is because the raw material of genes, nucleic acids, is relatively expensive for tiny prokaryotes to make, so there is strong selection at work to delete unnecessary sequences. In fact, the real challenge for viruses is discovering them in the first place. It is startling to think that all major virus discoveries, including HIV, weren't made by sequencing at all but by old methods such as magnifying them visually and looking at their morphology. Continuing advances in molecular technology have taught us the remarkable diversity of the virosphere, but this technology can only help us count the genes of something we already know exists.

Flourishing with Even Fewer

The number of genes we actually need for a healthy life is probably even fewer than the current estimate of 20,000 in our entire genome. One large study has reasonably extrapolated that the count of essential genes in human beings may be much lower. Researchers looked at thousands of healthy adults, looking for naturally occurring "knockouts," in which the function of particular genes is absent.[4] All our genes come in two copies—one from each parent. Usually, one active copy can compensate when the other is inactive, and it is difficult to find people with both copies inactive because inactivated genes are naturally rare.

Knockout genes are fairly easy to study with lab rats by using modern genetic engineering techniques to inactivate both copies of genes of our choice, or even to remove them altogether, and see what happens. But human studies require populations of people living in communities with 21st-century medical technologies and known pedigrees suited to the genetic and statistical analyses required. Icelanders are one useful population, and the British-Pakistani people of the already-cited knockout study are another. This research found over 700 genes that can be knocked out with no obvious health consequences. For instance, one surprising discovery was that the *PRDM9* gene, which plays a crucial role in the fertility of mice, can be knocked out in people with no ill effects.

So what genes do we need? We don't even know what a quarter of human genes actually do, and this is advanced compared with our knowledge of other species.[5]

Complexity Arises from the Very Simple

But whether the final number of human genes is 20,000 or 3,000 or something else, the point is that when it comes to understanding complexity, size really does not matter. We've known this for a long time in at least two contexts, and we are just beginning to understand the third.

Alan Turing, the mathematician and code breaker of World War II, established a theory of multicellular development. He studied simple mathematical models, now called "reaction-diffusion" processes, in which a small number of chemicals—just two in Turing's model—diffuse and react with each other. With simple rules governing their reactions, these models can reliably generate very complex, yet coherent, structures that are easily seen. So the biological structures of plants and animals do not require complex programming.

Similarly, it is obvious that the 100 trillion connections in the human brain, which are what really make us who we are, cannot possibly be genetically programmed individually. The recent breakthroughs in artificial intelligence are based on neural networks: these are computer models of the brain in which simple elements, corresponding to neurons, establish their own connections through interacting with the world. The results have been spectacular in applied areas such as handwriting recognition and medical diagnosis, and Google has invited the public to play games with and observe the dreams of its creations endowed with artificial intelligence.

Microbes Go beyond Basic

So it's clear that a single cell does not need to be especially complicated for large numbers of them to produce very complex outcomes. Hence, it should not come as a great surprise that human gene numbers may be on the same order as those of single-celled microbes like viruses and bacteria.

What does come as a surprise is the converse: that tiny microbes can have rich, complex lives. There is a growing field of study—dubbed "sociomicrobiology"—that examines the extraordinarily complex social lives of microbes, which stand up in comparison with our own. My own contributions to this area concern giving viruses their rightful place in this microscopic soap opera.

We have become aware in the last decade that microbes spend over 90 percent of their lives as biofilms, which may best be thought of as biological tissue. Indeed, many biofilms have systems of electrical communication between cells, as brain tissue does, making them a model for studying brain disorders such as migraine and epilepsy. Biofilms can also be thought of as "cities of microbes,"[6] and the integration of sociomicrobiology and medical research is making rapid progress in many areas, such as the treatment of cystic fibrosis. The social lives of microbes in these cities—complete with cooperation, conflict, truth, lies, and even suicide—is fast becoming a major study area in evolutionary biology in the 21st century.

Just as the biology of humans becomes starkly less outstanding than we had thought, the world of microbes gets far more interesting. And the number of genes doesn't seem to have anything to do with it.

Notes

1. Pertea, M., & Salzberg, S. L. (2010). Between a chicken and a grape: Estimating the number of human genes. *Genome Biology, 11*(5), 206. https://doi.org/10.1186/gb-2010-11-5-206.
2. D'Hont, A., Denoeud, F., Aury, J.-M., Baurens, F.-C., Carreel, F., Garsmeur, O., Noel, B., Bocs, S., Droc, G., Rouard, M., Da Silva, C., Jabbari, K., Cardi, C., Poulain, J., Souquet, M., Labadie, K., Jourda, C., Lengellé, J., Rodier-Goud, M., . . . Wincker, P. (2012). The banana (*Musa acuminata*) genome and the evolution of monocotyledonous plants. *Nature, 488*(7410), 213–217. https://doi.org/10.1038/nature11241.
3. Dagan, T., Roettger, M., Stucken, K., Landan, G., Koch, R., Major, P., Gould, S. B., Goremykin, V. V., Rippka, R., Tandeau de Marsac, N.,

Gugger, M., Lockhart, P. J., Allen, J. F., Brune, I., Maus, I., Pühler, A., & Martin, W. F. (2012). Genomes of stigonematalean cyanobacteria (subsection V) and the evolution of oxygenic photosynthesis from prokaryotes to plastids. *Genome Biology and Evolution, 5*(1), 31–44. https://doi.org/10.1093/gbe/evs117.

4. Narasimhan, V. M., Hunt, K. A., Mason, D., Baker, C. L., Karczewski, K. J., Barnes, M. R., Barnett, A. H., Bates, C., Bellary, S., Bockett, N. A., Giorda, K., Griffiths, C. J., Hemingway, H., Jia, Z., Kelly, M. A., Khawaja, H. A., Lek, M., McCarthy, S., McEachan, R., . . . van Heel, D. A. (2016). Health and population effects of rare gene knockouts in adult humans with related parents. *Science, 352*(6284), 474–477. https://doi.org/10.1126/science.aac8624.

5. Gollery, M., Harper, J., Cushman, J., Mittler, T., Girke, T., Zhu, J.-K., Bailey-Serres, J., & Mittler, R. (2006). What makes species unique? The contribution of proteins with obscure features. *Genome Biology, 7*(7). https://doi.org/10.1186/gb-2006-7-7-r57.

6. Watnick, P., & Kolter, R. (2000). Biofilm, city of microbes. *Journal of Bacteriology, 182*(10), 2675–2679. https://doi.org/10.1128/jb.182.10.2675-2679.2000.

Chapter 8

Everything You Wanted to Know about the First Cloned Mammal—Dolly the Sheep

GEORGE E. SEIDEL*

IT'S BEEN MORE THAN 25 years since scientists in Scotland told the world about Dolly the sheep, the first mammal successfully cloned from an adult body cell.[1] What was special about Dolly is that her "parents" were actually a single cell originating from the mammary tissue of an adult ewe. Dolly was an exact genetic copy of that sheep—a clone.

---

* *Editor's note*: Sadly, the author of this article, George E. Seidel, passed away in September 2021.

Dolly captured people's imaginations, but those of us in the field had seen her coming through previous research.[2] I've been working with mammalian embryos for over 40 years,[3] with some work in my lab specifically focusing on various methods of cloning cattle and other livestock species. In fact, one of the coauthors of the paper announcing Dolly worked in our laboratory for three years prior to going to Scotland to help create the famous clone.

Dolly was an important milestone, inspiring scientists to continue improving cloning technology as well as to pursue new concepts in stem cell research. The endgame was never meant to be armies of genetically identical livestock; rather, researchers continue to refine the techniques and combine them with other methods to turbocharge traditional animal breeding methods as well as gain insights into aging and disease.

Not the Usual Sperm + Egg

Dolly was a perfectly normal sheep that became the mother of numerous normal lambs. She lived to be six and a half years old, when she was eventually put down after a contagious disease had spread through her flock, infecting cloned and normally reproduced sheep alike. Her life wasn't unusual; it's her origin that made her unique.

Before the decades of experiments that led to Dolly, it was thought that normal animals could be produced only by fertilization of an egg by a sperm. That's how things naturally work. These germ cells are the only ones in the body that have their genetic material all jumbled up and in half the quantity of every other kind of cell. That way, when these so-called haploid cells come together at fertilization, they produce one cell with the full complement of DNA. Joined together, the cell is termed diploid, for twice or double. Two halves make a whole.

From that moment forward, nearly all cells in that body have the same genetic makeup. When the one-cell embryo duplicates its genetic material, both cells of the now two-cell embryo are genetically identical. When they in turn duplicate their genetic material, each cell at the four-cell stage is genetically identical. This pattern goes on such that each of

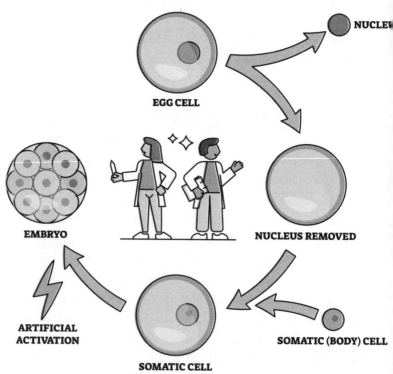

EGG CELL

NUCLE

NUCLEUS REMOVED

EMBRYO

ARTIFICIAL
ACTIVATION

SOMATIC (BODY) CELL

SOMATIC CELL
INTRODUCED INTO EGG CELL

the trillions of cells in an adult is genetically exactly the same—whether it's in a lung or a bone or the blood.

In contrast, Dolly was produced by what's called somatic cell nuclear transfer. In this process, researchers remove the genetic material from an egg and replace it with the nucleus of some other body cell. The resulting egg becomes a factory to produce an embryo that develops into an offspring. No sperm is in the picture; instead of half the genetic material coming from a sperm and half from an egg, it all comes from a single cell. It's diploid from the start.

Long Research Path to Dolly

Dolly was the culmination of hundreds of cloning experiments that, for example, showed diploid embryonic and fetal cells could be parents of offspring.[4] But there was no easy way to know all the characteristics of the animal that would result from a cloned embryo or fetus. Researchers could freeze a few of the cells of a 16-cell embryo, while going on to produce clones from the other cells; if a desirable animal was produced, they could thaw the frozen cells and make more copies. But this was impractical because of low success rates.

Dolly demonstrated that adult somatic cells also could be used as parents. Thus, one could know the characteristics of the animal being cloned. By my calculations, Dolly was the single success from 277 tries at somatic cell nuclear transfer. Sometimes the process of cloning by somatic cell nuclear transfer still produces abnormal embryos, most of which die. But the

Somatic cell nuclear transfer. In somatic cell nuclear transfer, all the DNA comes from a single adult cell. *VectorMine / iStock via Getty Images Plus* process has greatly improved; it's highly variable, though, depending on the cell type used and the species. These days most cloning is done using cells obtained from biopsied skin.

More Than Genes Can Affect a Clone

Genetics is only part of the story. Even though clones are genetically identical, their phenotypes—the characteristics they express—will be different. It's the same with naturally occurring identical twins: they share all their genes, but they're not exactly alike, especially if reared in different settings.

Environment plays a huge role for some characteristics. Food availability can influence weight. Diseases can stunt growth. These effects of lifestyle, nutrition, or disease can influence which genes are turned on or off in an individual—so-called epigenetic effects. Even when all the genetic material is the same in two identical clones, they may not be expressing all the same genes.

Consider the practice of cloning winning racehorses. Clones of winners sometimes will be winners too, but most of the time they're not. This is because winners are outliers: they must have the right genetics, yet they also need the right epigenetics and the right environment to reach that winning potential. For example, one can never perfectly duplicate the uterine conditions a winning racehorse experienced when it was a developing fetus. Thus, cloning champions usually leads to disappointment. On the other hand, cloning a stallion that sires a high proportion of race-winning horses will reliably result in a clone that similarly sires winners. This is a genetic rather than a phenotypic situation.

Even though the genetics are reliable, there are aspects of the cloning procedure that mean the epigenetics and environment are suboptimal. For example, sperm have elegant ways of activating the eggs they fertilize, which will die unless activated properly; with cloning, activation usually is accomplished by a strong electric shock. Many of the steps of cloning and subsequent embryonic development are done in test tubes in incubators. These conditions are not perfect substitutes for the female reproductive tract, where fertilization and early embryonic development normally occur.

Sometimes abnormal fetuses develop to term, resulting in abnormalities at birth. The most striking abnormal phenotype of some clones is termed "large offspring syndrome," in which calves or lambs are 30 to 40 percent larger than normal, resulting in a difficult birth. The problems stem from an abnormal placenta. At birth, these clones are genetically normal but oversize, and they tend to be hyperinsulinemic and hypoglycemic. (The conditions normalize over time once the offspring is no longer influenced by the abnormal placenta.) Recent improvements in cloning procedures have greatly reduced these abnormalities, which also occur with natural reproduction but at a much lower incidence.

Onward with Cloning

Many thousands of cloned mammals have been produced of nearly two dozen species. Very few of these concern practical applications, such as cloning a famous Angus bull named Final Answer (that died at an old age in 2014), in order to produce more high-quality cattle with his clone's sperm. The driving force for producing Dolly was not to create genetically identical animals. Rather, researchers wanted to combine cloning techniques with other methods in order to genetically change animals efficiently—much quicker than happens with traditional animal breeding methods, which take decades to make changes in populations of species such as cattle.

One example is introducing the polled (no horns) gene into dairy cattle,[5] thus eliminating the need for the painful procedure of dehorning. An even more striking application has been to produce a strain of pigs that is incapable of being infected by the highly contagious virus that causes the debilitating PRRS, or porcine reproductive and respiratory syndrome.[6] Researchers have even made cattle that cannot develop mad cow disease.[7] For each of these breeding procedures, somatic cell nuclear transplantation is an essential step.

To date, the most valuable contribution of these experiments in somatic cell nuclear transplantation has been the scientific information and insights gained. They've enhanced our understanding of normal and abnormal embryonic development, including aspects of aging. This information is already helping to reduce birth defects, improve methods of circumventing infertility, develop tools to fight certain cancers, and even decrease some of the negative consequences of aging— in livestock and even in people. More than two decades since Dolly, important applications are still evolving.

Notes

1. Wilmut, I., Schnieke, A. E., McWhir, J., Kind, A. J., & Campbell, K. H. (1997). Viable offspring derived from fetal and adult mammalian cells. *Nature*, *385*(6619), 810–813. https://doi.org/10.1038/385810a0.
2. Campbell, K. H., McWhir, J., Ritchie, W. A., & Wilmut, I. (1996). Sheep cloned by nuclear transfer from a cultured cell line. *Nature*, *380*(6569), 64–66. https://doi.org/10.1038/380064a0.
3. Seidel, G. E. (1983). Production of genetically identical sets of mammals: Cloning? *Journal of Experimental Zoology*, *228*(2), 347–354. https://doi.org/10.1002/jez.1402280217.
4. Seidel, G. E. (2015). Lessons from reproductive technology research. *Annual Review of Animal Biosciences*, *3*(1), 467–487. https://doi.org/10.1146/annurev-animal-031412-103709.

5. Carlson, D. F., Lancto, C. A., Zang, B., Kim, E.-S., Walton, M., Oldeschulte, D., Seabury, C., Sonstegard, T. S., & Fahrenkrug, S. C. (2016). Production of hornless dairy cattle from genome-edited cell lines. *Nature Biotechnology, 34*(5), 479–481. https://doi.org/10.1038/nbt.3560.

6. Whitworth, K. M., Rowland, R. R., Ewen, C. L., Trible, B. R., Kerrigan, M. A., Cino-Ozuna, A. G., Samuel, M. S., Lightner, J. E., McLaren, D. G., Mileham, A. J., Wells, K. D., & Prather, R. S. (2016). Gene-edited pigs are protected from porcine reproductive and respiratory syndrome virus. *Nature Biotechnology, 34*(1), 20–22. https://doi.org/10.1038/nbt.3434.

7. Richt, J. A., Kasinathan, P., Hamir, A. N., Castilla, J., Sathiyaseelan, T., Vargas, F., Sathiyaseelan, J., Wu, H., Matsushita, H., Koster, J., Kato, S., Ishida, I., Soto, C., Robl, J. M., & Kuroiwa, Y. (2006). Production of cattle lacking prion protein. *Nature Biotechnology, 25*(1), 132–138. https://doi.org/10.1038/nbt1271.

Chapter 9

From CRISPR to Glowing Proteins to Optogenetics—Scientists' Most Powerful Technologies Have Been Borrowed from Nature

MARC ZIMMER

WATSON AND CRICK, Schrödinger and Einstein all made theoretical breakthroughs that have changed the world's understanding of science.

Today, big, game-changing ideas are less common. New and improved techniques are the driving force behind modern scientific research and discoveries. They allow scientists—including chemists like me—to do our experiments faster than before, and they shine light on areas of science hidden to our predecessors.

Three cutting-edge techniques—the gene-editing tool CRISPR, fluorescent proteins, and optogenetics—were all inspired by nature. Biomolecular tools that have worked for bacteria, jellyfish, and algae for millions of years are now being used in medicine and biological research. Directly or indirectly, they will change people's lives.

Bacterial Defense Systems as Genetic Editors

Bacteria and viruses battle themselves and one another. They are at constant biochemical war, competing for scarce resources.[1] One of the weapons that bacteria have in their arsenal is the CRISPR-Cas system. It is a genetic library consisting of short repeats of DNA gathered over time from hostile viruses, paired with a protein called Cas that can cut viral DNA as though with scissors. In the natural world, when bacteria are attacked by viruses whose DNA has been stored in the CRISPR archive, the CRISPR-Cas system hunts down, cuts, and destroys the viral DNA.

Scientists have repurposed these weapons for their own use, with groundbreaking effect. Jennifer Doudna, a biochemist at the University of California, Berkeley, and French microbiologist Emmanuelle Charpentier shared the 2020 Nobel Prize in Chemistry for the development of CRISPR-Cas as a gene-editing technique.

The Human Genome Project has provided a nearly complete genetic sequence for humans and has given scientists a template to sequence the genomes of all other organisms. Before CRISPR-Cas, however, we researchers didn't have

tools for easily accessing and editing the genes in living organisms. Today, thanks to CRISPR-Cas, lab work that used to take months and years and cost hundreds of thousands of dollars can now be done in less than a week for just a few hundred dollars.

There are more than 10,000 genetic disorders caused by mutations that occur on only one gene, the so-called single-gene disorders. They affect millions of people. Sickle cell anemia, cystic fibrosis, and Huntington's disease are among the most well known of them. These are all obvious targets for CRISPR therapy because it is much simpler to fix or replace just one defective gene than it is to correct errors on multiple genes.

For example, in preclinical studies, researchers injected an encapsulated CRISPR system into patients born with a rare genetic disease, transthyretin amyloidosis, which causes fatal nerve and heart conditions. Preliminary results demonstrated that CRISPR-Cas can be injected directly into patients in such

French microbiologist Emmanuelle Charpentier (*left*) and US biochemist Jennifer Doudna shared the 2020 Nobel Prize in Chemistry for developing the CRISPR-Cas technique of gene editing. *Miguel Riopa/AFP via Getty Images*

a way that it can find and edit the faulty genes associated with a disease such as transthyretin amyloidosis. In the six patients included in this landmark work, the encapsuled CRISPR-Cas mini-missiles reached their target genes and did their job, causing a significant drop in a misfolded protein associated with the disease.[2]

Jellyfish Light Up the Microscopic World

The crystal jellyfish, *Aequorea victoria*, which drifts aimlessly in the northern Pacific, has no brain, no anus, and no poisonous stingers. It is an unlikely candidate to ignite a revolution in biotechnology. Yet on the periphery of its umbrella, it has about 300 photo-organs that give off pinpricks of green light that have changed the way science is conducted.

This bioluminescent light in the jellyfish stems from a luminescent protein called aequorin and a fluorescent protein called green fluorescent protein, or GFP. In modern biotechnology GFP acts as a molecular lightbulb that can be fused to other proteins, allowing researchers to track them and see when and where proteins are being made in the cells of living organisms. Fluorescent protein technology is used in thousands of labs every day and has resulted in the awarding of two Nobel Prizes, one in 2008 and the other in 2014. And fluorescent proteins have now been found in many more species.[3]

This technology proved its utility once again when researchers created genetically modified COVID-19 viruses that express GFP.[4] The resulting fluorescence makes it possible to follow the path of the viruses as they enter the respiratory system and bind to surface cells with hairlike structures.

Algae Let Us Play the Brain, Neuron by Neuron

When algae, which depend on sunlight for growth, are placed in a large aquarium in a darkened room, they swim around without direction. But if a lamp is turned on, the algae will swim toward the light. The single-celled flagellates—so named for the whiplike appendages they use to move around—don't have eyes. Instead, they have a structure called an eyespot that distinguishes between light and dark. The eyespot is studded with light-sensitive proteins called channelrhodopsins.

In the early 2000s, researchers discovered that when they genetically inserted these channelrhodopsins into the nerve cells of any organism, illuminating the channelrhodopsins with blue light caused neurons to fire. This technique, known as optogenetics, involves inserting the algae gene that makes channelrhodopsin into neurons. When a pinpoint beam of blue light is shined on these neurons, the channelrhodopsins open up, calcium ions flood through the neurons, and the neurons fire.

Using this tool, scientists can stimulate groups of neurons selectively and repeatedly, thereby gaining a more precise understanding of which neurons to target in treating specific disorders and diseases. Optogenetics might hold the key to treating debilitating and deadly brain diseases, such as Alzheimer's and Parkinson's.

But optogenetics isn't only useful for understanding the brain. Researchers have used optogenetic techniques to partially reverse blindness,[5] and they have found promising results in clinical trials using optogenetics on patients with retinitis pigmentosa, a group of genetic disorders that break down retinal cells. And in mouse studies, the technique has been used to manipulate heartbeat and to regulate the bowel movements of constipated mice.

What Else Lies in Nature's Toolbox?

What undiscovered techniques does nature still hold for us?

According to a 2018 study, people represent just 0.01 percent of all living things by mass but have caused the loss of 83 percent of all wild mammals and half of all plants in our brief time on Earth. By annihilating nature, humankind might be losing out on new, powerful, and life-altering techniques before having even imagined them.

After all, no one could have foreseen that the discovery of three groundbreaking processes derived from nature would change the way science is done.

Notes

1. Granato, E. T., Meiller-Legrand, T. A., & Foster, K. R. (2019). The evolution and ecology of bacterial warfare. *Current Biology, 29*(11). https://doi.org/10.1016/j.cub.2019.04.024.
2. Gillmore, J. D., Gane, E., Taubel, J., Kao, J., Fontana, M., Maitland, M. L., Seitzer, J., O'Connell, D., Walsh, K. R., Wood, K., Phillips, J., Xu, Y., Amaral, A., Boyd, A. P., Cehelsky, J. E., McKee, M. D., Schiermeier, A., Harari, O., Murphy, A., . . . Lebwohl, D. (2021). CRISPR-Cas9 in vivo gene editing for transthyretin amyloidosis. *New England Journal of Medicine, 385,* 493–502. https://doi.org/10.1056/nejmoa2107454.

3. FPbase, a fluorescent protein database, aggregates these findings: https://www.fpbase.org/about/.

4. Hou, Y. J., Okuda, K., Edwards, C. E., Martinez, D. R., Asakura, T., Dinnon, K. H., Kato, T., Lee, R. E., Yount, B. L., Mascenik, T. M., Chen, G., Olivier, K. N., Ghio, A., Tse, L. V., Leist, S. R., Gralinski, L. E., Schäfer, A., Dang, H., Gilmore, R., . . . Baric, R. S. (2020). SARS-CoV-2 reverse genetics reveals a variable infection gradient in the respiratory tract. *Cell, 182*(2), P429–446. https://doi.org/10.1016/j.cell.2020.05.042.

5. Sahel, J.-A., Boulanger-Scemama, E., Pagot, C., Arleo, A., Galluppi, F., Martel, J. N., Esposti, S. D., Delaux, A., de Saint Aubert, J.-B., de Montleau, C., Gutman, E., Audo, I., Duebel, J., Picaud, S., Dalkara, D., Blouin, L., Taiel, M., & Roska, B. (2021). Partial recovery of visual function in a blind patient after optogenetic therapy. *Nature Medicine, 27*(7), 1223–1229. https://doi.org/10.1038/s41591-021-01351-4.

Part II.

Biotechnology, Food, and the Environment

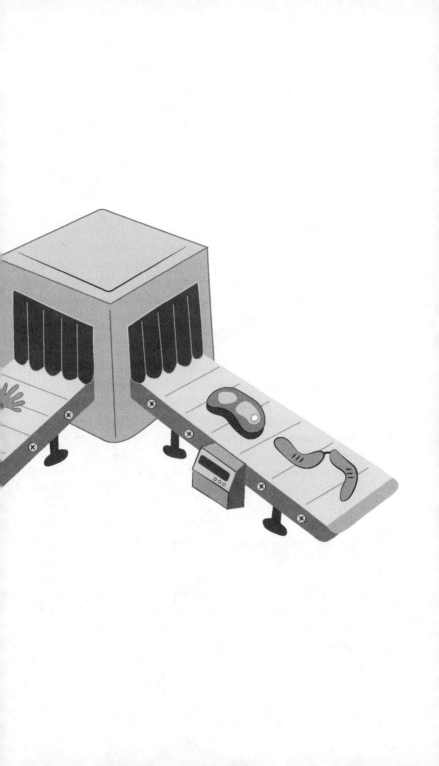

The use of genetically modified organisms, or GMOs, in food production has always been controversial. Even though the World Health Organization and the National Academies of the United States, of the United Kingdom, and of France have all released statements proclaiming there to be no higher risk to humans from consuming GMOs than from consuming other foods, about half of the American public still believes that GMO foods are worse for you, and that number is higher in Europe.[1] There is no doubt that many GMO foods do not deliver on what they promise, but there is a huge potential for CRISPR'd crops to use less water and pesticides, while at the same time being more nutritious. That is why a contingent of researchers is campaigning for some GMOs to be labeled "organic" produce.

Besides producing more sustainable foods, biotechnologies also offer many opportunities for a cleaner environment. There are efforts to reduce disease and pesticide use, for example, by designing *Aedes* mosquitoes with CRISPR gene drives that will wipe out, in a targeted fashion, their dengue- and Zika-carrying fellow *Aedes* mosquitoes, and people are using synthetic biology to genetically modify bacteria to eat pollutants found in mining and oil sands waste. Raising meat accounts for 57 percent of all greenhouse gasses released by food production.[2] Thanks to some ingenious non-GMO plant-based meat substitutes, we are at the point where plant-based fast-food burgers are indistinguishable from their traditional meat equivalents.

The chapters in part II survey the potential of genetic engineering in food production and in environmental protection and discuss some of the regulatory questions that this new technology raises. They highlight how controversial and complicated the field is and how it is pulled in different directions by public opinion, expert knowledge, regulations, and hype. These chapters were selected to give an overview of GMOs; the selections lean toward science in their orientation, but some historical and ethical perspectives round out the group.

Notes

1. *The new food fights: U.S. public divides over food science.* Section 3. Public opinion about genetically modified foods and trust in scientists connected with these foods. (2016, December 16). Pew Research Center. https://www.pewresearch.org/science/2016/12/01/public -opinion-about-genetically-modified-foods-and-trust-in-scientists -connected-with-these-foods/.
2. Milman, O. (2021, September 13). Meat accounts for nearly 60% of all greenhouse gases from food production. *The Guardian.* https://www .theguardian.com/environment/2021/sep/13/meat-greenhouses -gases-food-production-study.

Chapter 10

What Is Bioengineered Food?
An Agriculture Expert Explains

KATHLEEN MERRIGAN

THE US DEPARTMENT OF AGRICULTURE defines bioengineered foods "as those that contain detectable genetic material that has been modified through certain lab techniques and cannot be created through conventional breeding or found in nature."[1] If that definition sounds familiar, it is because that's essentially how genetically modified organisms, or GMOs, are defined. On January 1, 2022, the USDA implemented a new disclosure standard for bioengineered food. Shoppers are seeing labels on food products with the term "bioengineered" or the phrase "derived from bioengineering" printed on a green seal with an icon of the sun shining down on cropland.

More than 90 percent of US-grown corn, soybeans, and sugar beets are genetically modified.[2] This means that many processed foods containing high-fructose corn syrup, beet sugar, or soy protein may fall under the new disclosure standard. Other whole foods on the USDA's list of bioengineered foods, such as certain types of eggplant, potatoes, and apples, may have to carry labels as well.

Disclosure Debates

Food manufacturers have historically opposed labeling. They argue that it misleads consumers into thinking that bioengineered foods are unsafe. Countless studies, along with the USDA and the World Health Organization, have concluded that eating genetically modified foods does not pose health risks.[3]

Many consumers, though, have demanded labels that let them know whether foods contain genetically modified material. In 2014 Vermont enacted a strict law mandating GMO food labeling. Fearing a checkerboard of state laws and regulations, food manufacturers lobbied successfully for a federal disclosure law to preempt other states from doing the same. When it introduced the label, the United States joined 64 other countries that require some sort of labeling.[4]

Consumer and right-to-know advocates are not happy with the new federal disclosure standard, however. The Center for Food Safety, the lead organization representing a coalition of food-labeling nonprofits and retailers, has filed suit against the USDA, arguing that the standard not only fails to use common language but is also deceptive and discriminatory.

According to this view, the standard is deceptive because loopholes exclude many bioengineered foods from mandatory disclosure, which critics say is inconsistent with consumer expectations. If the genetic material is undetectable or less than 5 percent of the finished product, no disclosure is required. As a result, many highly refined products—for example, sugar or oil made from a bioengineered crop—may be excluded from labeling requirements. Bioengineered foods served in restaurants, cafeterias, and transport systems, including food trucks, are also excluded. And the standard excludes meat, poultry, and eggs, as well as products that list those foods as either their first ingredient or their second ingredient after water, stock, or both. It takes a 43-minute USDA webinar to explain what's in and what's out under this new disclosure standard.

Critics say the standard is discriminatory because it gives food manufacturers disclosure options that can substitute for the green bioengineered seal. They include listing a phone number to call or text for information or a QR code that's scannable with a smartphone. But critics point out that many people in the United States lack access to a smartphone, particularly those over 65 and those earning less than $30,000 annually.[5]

In my view, consumers who want to avoid bioengineered foods may best be served by buying products that are certified organic, a designation that prohibits genetically modified ingredients. Or they can search for the voluntary Non-GMO Project Verified label, which features a butterfly.[6] It was launched in 2010 and appears on tens of thousands of grocery items. Both labels indicate that a third-party inspector verified that the non-GMO standard has been met.

The federal labeling standard came to market with little fanfare—probably because neither side in the conflict over the genetic modification of food sees it as a win.

Notes

1. *BE disclosure.* (n.d.). United States Department of Agriculture. https://www.ams.usda.gov/rules-regulations/be.
2. National Agricultural Statistics Service (2021, June 30). *Acreage.* United States Department of Agriculture. https://downloads.usda .library.cornell.edu/usda-esmis/files/jO98zbO9z/00000x092 /kw52k657g/acrg0621.pdf.
3. National Academies of Sciences, Engineering, and Medicine. (2016). *Genetically engineered crops: Experiences and prospects.* National Academies Press.
4. *Genetically engineered food labeling laws.* (n.d.). Center for Food Safety. https://www.centerforfoodsafety.org/ge-map/.
5. Mobile fact sheet. (2021, April 7). Pew Research Center. https://www .pewresearch.org/internet/fact-sheet/mobile/.
6. See Non-GMO Project. https://www.nongmoproject.org/.

Chapter 11

Organic Farming with Gene Editing: An Oxymoron or a Tool for Sustainable Agriculture?

REBECCA MACKELPRANG

A UNIVERSITY OF CALIFORNIA, BERKELEY, professor stands at the front of the room, delivering her invited talk about the potential of genetic engineering. Her audience, full of organic farming advocates, listens uneasily. She notices a man get up from his seat and move toward the front of the room. Confused, the speaker pauses mid-sentence as she watches him bend over, reach for the power cord, and unplug the projector. The room darkens and silence falls. So much for listening to the ideas of others.

Many organic advocates claim that genetically engineered crops are harmful to human health, the environment, and the farmers who work with them. Biotechnology advocates fire back that genetically engineered crops are safe,[1] reduce insecticide use, and can enable subsistence farmers in developing countries to produce a more consistent crop yield for themselves and their families. Sides are still being drawn about whether CRISPR, a gene editing technology, is really just "GMO 2.0" or is instead a helpful new tool to speed up the plant breeding process.[2]

I am trained as a plant molecular biologist and appreciate the awesome potential of both CRISPR and genetic engineering technologies. But I don't believe that pits me against the goals of organic agriculture. In fact, biotechnology can help meet these goals. And while rehashing the arguments about genetic engineering seems counterproductive, genome editing may draw both sides to the table for a healthy conversation. To understand why, it's worth digging into the differences between genome editing with CRISPR and genetic engineering.

How Do Genetic Engineering, CRISPR, and Mutation Breeding Differ?

Opponents argue that CRISPR is a sneaky way to trick the public into eating genetically engineered foods. It is tempting to toss CRISPR and genetic engineering into the same bucket. But genetic engineering and CRISPR, if referred to categorically, are too broad to convey what is happening on the genetic level, so let's look closer.

In one type of genetic engineering, a gene from an unrelated organism can be introduced into a plant's genome. For example, a significant percentage of the eggplant grown in Bangladesh includes a gene from a common bacterium.[3] This gene makes a protein called Bt that is harmful to insects. With that gene inside the eggplant's DNA, the plant becomes lethal to eggplant-eating insects and so decreases the need for insecticides. Bt is safe for humans. As a comparison, consider how chocolate makes dogs sick but doesn't affect us.

Another type of genetic engineering can move a gene from one variety of a plant species into another variety of that same species. For example, researchers identified a gene in wild apple trees that makes them resistant to fire blight. They moved that gene into the Gala Galaxy apple to make it resistant to disease. However, as of early 2022, this new apple variety has not been commercialized.

Scientists are unable to direct where in the genome a gene is inserted with traditional genetic engineering, although they use DNA sequencing to identify the location after the fact. In contrast, CRISPR is a tool of precision.

Just like the "find" function of a word processor lets a user jump to a particular word in a document, the CRISPR molecular machinery finds a specific spot in a genome. It cuts both strands of DNA at that location. Because cut DNA is problematic for the cell, it quickly deploys a "repair team" to mend the break. There are two pathways for repairing the DNA. In one, which I call "CRISPR for modification," a new gene can be inserted to link the cut ends together, like pasting a new sentence into a document.

In "CRISPR for mutation," the cell's repair team tries to glue the cut DNA strands back together again. Sometimes the repair team makes an error, thereby introducing a small DNA change called a mutation. Scientists can even help direct this repair team to make desired "errors" at the site that was cut. This technique can be used to tweak the gene's behavior inside the plant. It can also be used to inactivate genes inside the plant that, for example, are detrimental to plant survival, like a gene that increases susceptibility to fungal infections.[4]

Mutation breeding, which in my opinion is also a type of biotechnology, is already used in organic food production. In mutation breeding, radiation or chemicals are used to randomly make mutations in the DNA of hundreds or thousands of seeds, which are then grown in the field. Breeders inspect fields for plants with a desired trait such as disease resistance or increased yield. Thousands of new crop varieties have been created and commercialized through this process, including everything from varieties of quinoa to varieties of grapefruit.[5] Mutation breeding is considered a traditional breeding technique and thus is not an "excluded method" for organic farming in the United States.

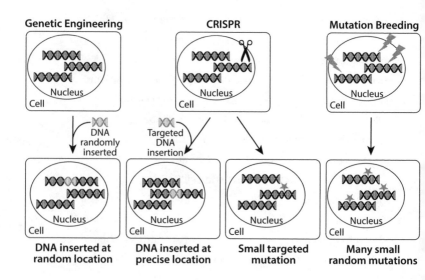

CRISPR for mutation is more like mutation breeding than like genetic engineering. It creates similar end products as mutation breeding, but it removes the randomness. It does not introduce new DNA. It is a controlled and predictable technique for generating helpful new plant varieties capable of resisting disease or weathering adverse climatic conditions.

Opportunity Lost—Learning from Genetic Engineering

Most genetically engineered traits that are commercialized confer herbicide tolerance or insect resistance in corn, soybean, or cotton. Yet many other engineered crops exist. While a few are grown in the field, most sit all but forgotten in dark corners of research labs because of the prohibitive expense of clearing regulatory hurdles. If the regulatory climate and public perception would allow it, crops with valuable traits could be produced by CRISPR and become common in our crops and on our tables.

For example, my former advisor at UC Berkeley developed, with colleagues, a hypoallergenic variety of wheat. Seeds for this wheat are held captive in envelopes in the basement of her building, untouched for years. A tomato that uses a sweet

In traditional genetic engineering, a new gene is added to a random location in an organism's genome. CRISPR for modification also allows a new gene to be added, but it introduces the new gene at a specific location in the genome. CRISPR for mutation does not add new DNA. Rather, it makes a small DNA change at a precise location. Older forms of mutation breeding for plants use chemicals or radiation (symbolized by lightning bolts) to induce small, randomly located mutations in the genomes of seeds. The resulting plants are screened for desirable phenotypes resulting from the induced mutations. *Rebecca Mackelprang, CC BY–SA* pepper gene to defend against a bacterial disease, decreasing the need for copper-based pesticide, has struggled to secure funding to move forward. Carrot, cassava, lettuce, potato, and more have been engineered for increased nutritional value. These varieties demonstrate the creativity and expertise of researchers in bringing beneficial new traits to life. Why, then, can't I buy bread made with hypoallergenic wheat at the grocery store?

Loosening the Grip of Big Agriculture

Research and development for a new genetically engineered crop costs around 100 million US dollars at large seed companies. Clearing the regulatory hurdles laid out by the US Department of Agriculture, the Environmental Protection Agency, and/or the Food and Drug Administration (depending on the engineered trait) has historically taken between five and seven years and an additional $35 million. Regulation is important, and genetically engineered products should be carefully evaluated. But the great expense has left only large corporations with substantial capital to compete in this arena, while shutting out small companies, academic researchers, and nongovernmental organizations. To recoup their $135 million investment in crop commercialization, companies develop products to satisfy the biggest markets of seed buyers: growers of corn, soybean, sugar beet, and cotton.

The costs of research and development are far lower with CRISPR owing to its precision and predictability, and in the United States, plants developed using "CRISPR for mutation" are exempt from regulation by the USDA.[6] As a result, CRISPR may escape the dominant financial grasp of large seed companies. Academics, small companies, and researchers working for nongovernmental organizations may see hard work and intellectual capital yield beneficial genome-edited products that are not forever relegated to the basements of research buildings.

Common Ground: CRISPR for Sustainability

In the years since the genome editing capabilities of CRISPR were unlocked, academics, startups, and established corporations have announced new agricultural products in the pipeline that use this technology. Some of these focus on traits for consumer health, such as low-gluten or gluten-free wheat for people with celiac disease. Others, such as non-browning mushrooms, can decrease food waste.

As the climate continues to change faster than evolution can keep up, rapid breeding for tolerance to drought and salinity is taking place for tomatoes, corn, rice, and many other foods. Tomatoes resistant to powdery mildew could save billions of dollars and eliminate the need for spraying fungicides. A tomato plant that flowers and makes fruit early could be used in northern latitudes with long days and short growing seasons, which will become more important as the climate changes.

Differing Views from Organic Farmers

The US National Organic Standards Board (NOSB) has voted to exclude all genome-edited crops from organic certification.[7] But in my view, the NOSB should reconsider.

Some of the organic growers I interviewed agreed. "I see circumstances under which it could be useful for shortcutting a process that for traditional breeding might take many plant generations," said Tom Willey, an organic farmer emeritus from California. The disruption of natural ecosystems is a major challenge to agriculture, Willey told me; and while the problem cannot be wholly addressed by genome editing, it could lend an opportunity to "reach back into genomes of the wild ancestors of crop species to recapture genetic material" that has been lost through millennia of breeding for high yields. Breeders have successfully used traditional breeding to reintroduce such diversity, but given "the urgency posed by climate change, we might wisely employ CRISPR to accelerate such work," Willey concluded.

Bill Tracy, an organic corn breeder and professor at the University of Wisconsin–Madison, said, "Many CRISPR-induced changes that could happen in nature could have benefits to all kinds of farmers." But the NOSB has already voted on the issue, and the rules are unlikely to change without significant pressure. "It's a question of what social activity could move the needle on that," Tracy concluded.

People on all sides of biotechnology debates want to maximize beneficial outcomes for humanity and the environment. Collaborative problem solving by organic (and conventional) growers, specialists in sustainable agriculture, biotechnologists, and policy makers will yield greater progress than will individual groups acting alone and dismissing one another. The barriers to this collaboration may seem large, but they are of our own making. I hope that more people in the future will leave the projector running and join in the discussion on genetically engineered crops.

Notes

1. National Academies of Sciences, Engineering, and Medicine. (2016). *Genetically engineered crops: Experiences and prospects.* National Academies Press. https://doi.org/10.17226/23395.

2. Scheben, A., & Edwards, D. (2018). Towards a more predictable plant breeding pipeline with CRISPR/Cas-induced allelic series to optimize quantitative and qualitative traits. *Current Opinion in Plant Biology*, *45*, 218–225. https://doi.org/10.1016/j.pbi.2018.04.013.

3. Shelton, A. M., Hossain, M. J., Paranjape, V., Azad, A. K., Rahman, M. L., Khan, A. S. M. M. R., Prodhan, M. Z. H., Rashid, M. A., Majumder, R., Hossain, M. A. and Hussain, S. S. (2018). Bt eggplant project in Bangladesh: History, present status, and future direction. *Frontiers in Bioengineering and Biotechnology*, article 106. https://doi.org/10.3389/fbioe.2018.00106.

4. Wang, Y., Cheng, X., Shan, Q., Zhang, Y., Liu, J., Gao, C., & Qiu, J.-L. (2014). Simultaneous editing of three homoeoalleles in hexaploid bread wheat confers heritable resistance to powdery mildew. *Nature Biotechnology, 32*(9), 947–951. https://doi.org/10.1038/nbt.2969.

5. Ahloowalia, B. S., Maluszynski, M., & Nichterlein, K. (2004). Global impact of mutation-derived varieties. *Euphytica, 135*(2), 187–204. https://doi.org/10.1023/b:euph.0000014914.85465.4f.

6. *Regulatory exemptions.* (n.d.). USDA Animal and Plant Health Inspection Service. https://www.aphis.usda.gov/aphis/ourfocus/biotechnology/permits-notifications-petitions/exemptions.

7. *Formal recommendation. From the National Organic Standards Board to the National Organic Program.* (n.d.). https://www.ams.usda.gov/sites/default/files/media/MSExcludedMethodsApr2019FinalRec.pdf.

Chapter 12

How We Got to Now: Why the US and Europe Went Different Ways on GMOs

PAUL B. THOMPSON

THERE IS A MYTH THAT CIRCULATES on both sides of the Atlantic: Americans accepted genetically modified organisms (GMOs) in their food supply without question, while the more cautious Europeans rejected them. This overlooks the fact that GMOs went through a period of significant controversy in the United States during the early years, starting in the 1980s.

A boomerang effect was felt in the United States around 2010 in reaction to a rise in consumer concern, state-based initiatives for labeling, and the emergence of "GMO-free" claims on a growing number of products marketed in the

United States. Congress enacted the National Bioengineered Food Disclosure Standard in 2016. In Europe, meanwhile, the controversy seems to have never subsided. Despite signs that the resistance of many European consumers is waning, the European Union has thus far applied much more restrictive regulation and labeling requirements to gene-edited foods.

Why have EU and US consumers and policy makers taken different routes? A look at the early days of GMOs helps explain why.

An Uproar over Dairy Cows

The first two genetically engineered food products in the United States were recombinant chymosin, or rennet (an enzyme used in cheese production), and recombinant bovine somatotrophin (BST), a growth hormone used to extend the lactation cycle in dairy cows. Both are produced in a genetically engineered microbe in much the same manner as many drugs. Recombinant rennet was accepted without a whisper in both the United States and Europe. Recombinant BST, on the other hand, caused an uproar.

It began in 1985 when economists predicted that recombinant BST (rBST) would lead to concentration in the dairy industry. The US dairy industry was already starting to consolidate on account of computerized record keeping, herd management, and the control of milking equipment. Worries that small dairies across the United States would go bankrupt worsened as the industry transitioned to milking not dozens or hundreds of cows at a facility but thousands of them thanks to the longer lactation cycle.

Consequently, rBST went through an extraordinarily long and drawn-out approval process by the Food and Drug

Administration (FDA) and was in fact withheld from the market after it was approved by a highly unusual act of Congress. The special review mandated by this act concurred with the FDA's assessment of rBST's safety and further stated that the US government had never before regulated a novel technology in light of predicted socioeconomic consequences. The moratorium on rBST was allowed to expire in the early years of the Clinton administration in the early 1990s, thus allowing rBST to go on the market.

This did not end the controversy, however. There were numerous attempts to promote labels for "rBST-free" milk, especially in New England, where people love their small dairies. And, in general, there is a tendency for any food-related claim to be regarded as a health claim by a subset of consumers. The FDA judged the rBST-free claim to be misleading since all milk contains BST, and it had already concluded that rBST milk was as safe as regular milk.

The FDA was quite aggressive in policing these claims. Ben & Jerry's ice cream, however, was one of the few companies willing to jump through all the hurdles to maintain its rBST-free label. The company added disclaimers saying that all milk has BST and that sourcing its milk from non-rBST dairies was found to have no health implications. By the time Ben & Jerry's added further required language, stating that the company couldn't be sure that all of its suppliers had done the same thing, the label that satisfied the FDA was a paragraph long.

Meanwhile, agencies in Canada and Europe ruled against rBST on grounds of animal health. Inducing higher milk production is accompanied by a statistical increase in the risk of mastitis.[1] The US, by contrast, was primed for a political environment that was pro-biotechnology and hostile to demands for regulation or labeling on any but the strictest of health-based claims.

Ethics

If the larger social context in agriculture was pro-biotech, this was certainly not true for a loose-knit coalition that was to prove its mettle in the years to come.

An almost forgotten document from 1990, *Biotechnology's Bitter Harvest* laid out a series of complaints.[2] Foremost among them were concerns about small-farm bankruptcies and concentration in agriculture and the tendency for US agricultural research to underfund and ignore environmentally friendly alternatives to large-scale monoculture, mechanization, and chemical inputs.

The authors of *Biotechnology's Bitter Harvest* predicted that genetic manipulation would follow this path, and they demanded that land-grant universities and the US Department of Agriculture (USDA) expand their portfolio to be more accommodating to production methods that we associate today with organic farming. It is at least arguable that had agricultural research institutions followed this advice, we would not see the extreme alienation and bifurcation between industrial and alternative agriculture that exists today.

There may also have been a brief moment when the biotechnology industry itself could have endorsed such a move. During the early 1990s, the nonprofit Keystone Center facilitated a series of "national conversations" on new genetic technologies that raised the ethical issues associated with both medical and food applications. I attended one of these sessions and read all the reports.

These efforts were a testament to the significant and growing dissatisfaction with mainstream agriculture, but the human medical questions were clearly the hulking gorilla in the room. The upshot of these talks was recognition that people want drugs that could be developed by manipulating genes, but they had ethical issues with applying genetic engineering to the human germ line.[3] Similar ethical concerns over the manipulation of food crops and especially food animals tapered off.

In any event, although concerns were being expressed, US regulatory agencies were reluctant to base their decisions on factors not clearly articulated by Congress in its authorizing legislation. US regulatory decisions can be and regularly are challenged in court. Although the internal discussions at the USDA, the FDA, and the Environmental Protection Agency (EPA) are not made public, we can presume that legal advisers at these agencies would have urged them to resist the pressure to consider anything but health and environmental impact, narrowly construed. The first genetically engineered crops were approved in the late 1990s, and by 2000 a large percentage of US corn and soybean farmers were growing GMO varieties.

Safety and Regulation

What about food safety? Understanding this part of the story requires a look at how food is regulated in the United States. The FDA has clear authority to regulate additives (like coloring agents or preservatives) and animal drugs (like rBST). Foods themselves, however, are not subject to any mandatory review under US law, and the FDA has long circulated a list of foods and food ingredients that are "generally recognized as safe," or GRAS. Food companies combining items on the GRAS list have a blanket endorsement from the FDA that shields them from lawsuits that might otherwise be brought under US liability law.

Meanwhile, dating back to the days of the first Bush administration, regulatory agencies had been directed to use existing laws to regulate biotechnology—that is, no mandatory review of GMO foods was required. This is a decision that remains controversial to this day.

The FDA eventually announced that it would treat any gene product, such as a protein or active agent produced by a genetic modification, as an additive, if the product was not itself from a source on the GRAS list; this adjustment to policy gave the FDA strong authority over truly novel introductions into food. But given that it had no authority to require regulatory review, the FDA was in the position of relying on biotechnology companies to report, voluntarily, what genes they had introduced into crops. The case for animals was different: all genetic modifications are regulated as animal drugs—a difference that may explain why no transgenic animals were approved for food use in the United States before 2015.

This approach has subsequently been called "substantial equivalence," which falls short of a regulatory approval, since the FDA only reviews data submitted by companies on the chemical composition of GRAS foods. GMOs do receive formal approval from the USDA and the EPA, but these agencies are reviewing environmental risk rather than food safety risk.

The approach of substantial equivalence has endured, in part because nothing has gone wrong (at least nothing we know of) and because the alternative is difficult to define. Natural variation in the chemical composition of virtually all common foods is quite large, which means that standard toxicological methods for testing the safety of whole foods are subject to many confounding variables.

Food safety experts are well aware that there are many ways in which ordinary plant breeding can produce unsafe whole foods. This would be especially true for foods such as tomatoes or potatoes, which are known to carry genes for potent toxins. However, there is no law in the United States that would require any whole food to be subjected to any regulatory review. The only protection that keeps toxic plants off the shelves of grocery stores in the United States is the professional ethic of plant breeders, reinforced by the fear of a product liability lawsuit. Indeed, the litigious nature of American society and the ready supply of trial lawyers eager to take a shot at any well-heeled company marketing an unsafe food is an important backstop that is often overlooked when comparing the US regulatory approach with the rest of the world.

US Biotech Goes to Europe

The development of GMO foods in Europe played out at the same time as the initial steps were taking place toward integrating national food safety systems into the European Food Safety Authority. It was politically contentious because nation-states were losing some of their influence over home-based regulation. For example, the *Reinheitsgebot*, or German beer purity laws, had virtually ensured that anything labeled beer had to have been produced in Germany. The economic interests of individual countries threatened by EU-wide food safety rules created a touchy political climate.

What is more, a series of high-profile food safety debacles undercut Europeans' confidence in the food and agricultural industry, as well as the regulatory science behind government-mandated assessments of food safety risk. Mad cow disease in the UK was the most prominent of these events,[4] while the radioactive contamination of European fields after the Chernobyl disaster led Europeans to be especially leery of bad scientific decisions made elsewhere.

The US biotechnology industry blustered its way into this already-touchy regulatory environment with GMO crops that it hoped to sell to European farmers. US biotech firms insisted that Europeans simply accept the safety assessments that had already been made by a trio of US regulatory agencies: the FDA, USDA and EPA. Needless to say, Europeans were not having any of it.

At the same time, European scientists themselves were moving into GMOs. A canned and labeled GMO tomato had been successfully test-marketed in the mid-1990s through a cooperative agreement between Sainsbury's, a major UK grocery chain, and the University of Nottingham. As news about the US biotechnology industry's attempt to force its way into European markets began to break, however, activists began campaigns against so-called Frankenfoods. Sainsbury's competitors began to advertise that their store brands were GMO-free, and Sainsbury dropped the experiment, saying, "our customers have indicated to us very clearly that they do not want genetically-modified ingredients."[5]

One lasting legacy of this episode is that European grocery stores are willing to compete against one another by making claims that impugn the safety of foods being sold by their competitors, while American grocery chains are generally not. The aggressive approach taken by the FDA against claims about rBST may well be a contributing factor to a legacy of

American stores accepting the safety of GMO products. And as the FDA has relaxed its efforts to police claims about the alleged health benefits of foods, the American food industry has shown signs of a willingness to attract customers by touting the attractiveness of organic or GMO-free foods.

A slightly more complete history would point to a number of other incidents that have led to the sharp division of opinion that exists today. The Flavr Savr tomato was the first genetically modified crop to be commercialized in 1994. Designed to stay ripe and firm longer, the product failed to meet the needs of the US tomato industry. There were other frictions and failures: ice-nucleating, or "Frostban," bacteria; StarLink corn; the Pusztai affair; African rejection of US food aid[6]—the list goes on.

At the same time, contemporary activists, who may have never heard of *Biotechnology's Bitter Harvest*, are now building steadily on the dissatisfaction expressed a quarter of a century ago to create an economically and politically vibrant "food movement" that wants nothing to do with biotechnology or genetically engineered foods.

Notes

1. Seegers, H., Fourichon, C., & Beaudeau, F. (2003). Production effects related to mastitis and mastitis economics in dairy cattle herds. *Veterinary Research, 34*(5), 475–491. https://doi.org/10.1051/vetres:2003027.
2. Goldburg, R., Rissler, J., Shand, H., & Hassebrook, C. (n.d.). *Biotechnology's bitter harvest: Herbicide-tolerant crops and the threat to sustainable agriculture.* Union of Concerned Scientists. https://blog.ucsusa.org/wp-content/uploads/2012/05/Biotechnologys-Bitter-Harvest.pdf.
3. Saey, T. H. (2021, November 22). *Editing human germline cells sparks ethics debate.* Science News. https://www.sciencenews.org/article/editing-human-germline-cells-sparks-ethics-debate.

4. *Timeline of mad cow disease outbreaks.* Center for Food Safety. (n.d.). https://www.centerforfoodsafety.org/issues/1040/mad-cow-disease/timeline-mad-cow-disease-outbreaks.
5. *UK Sainsbury's phase out GM food.* (1999, March 17). BBC News. http://news.bbc.co.uk/2/hi/uk_news/298229.stm.
6. Zerbe, N. (2004). Feeding the famine? American food aid and the GMO debate in Southern Africa. *Food Policy, 29*(6), 593–608. https://doi.org/10.1016/j.foodpol.2004.09.002.

Chapter 13

Can Genetic Engineering Save Disappearing Forests?

JASON A. DELBORNE

COMPARED TO GENE-EDITED BABIES in China and ambitious projects to rescue woolly mammoths from extinction, biotech trees might sound pretty tame. But releasing genetically engineered trees into forests to counter threats to forest health represents a new frontier in biotechnology. Even as the techniques of molecular biology have advanced, humans have not yet released a genetically engineered plant that is intended to spread and persist in an unmanaged environment. Biotech trees—genetically engineered or gene-edited—offer just that possibility.

One thing is clear: the threats facing our forests are many, and the health of these ecosystems is getting worse. A 2012 assessment by the US Forest Service estimated that nearly 7 percent of forests nationwide are in danger of losing at least a quarter of their tree vegetation by 2027.[1] This estimate may not sound too worrisome, but it is 40 percent higher than the previous estimate made just six years earlier.

In 2018, several US federal agencies and the US Endowment for Forestry and Communities requested that the National Academies of Sciences, Engineering, and Medicine form a committee to "examine the potential use of biotechnology to mitigate threats to forest tree health"; experts, including me, a social scientist focused on emerging biotechnologies, were asked to "identify the ecological, ethical, and social implications of deploying biotechnology in forests, and develop a research agenda to address knowledge gaps."[2]

Our committee members came from universities, federal agencies, and nongovernmental organizations and represented a range of disciplines: molecular biology, economics, forest ecology, law, tree breeding, ethics, population genetics, and sociology. All of these perspectives were important for considering the many potentials and challenges of using biotechnology to improve forest health.

A Crisis in US Forests

Forests today face higher temperatures and droughts and more pests than before. As goods and people move around the globe, more insects and pathogens hitchhike into our forests. We committee members focused on four case studies to illustrate the breadth of forest threats.

The emerald ash borer arrived from Asia and causes severe mortality in five species of ash trees. First detected on US soil in 2002, it had spread to 31 states as of May 2018. Whitebark pine, a keystone and foundational species in high elevations of the United States and Canada, is under attack by the native mountain pine beetle and an introduced fungus. Over half of whitebark pine trees in the northern United States and Canada have died.

Poplar trees are important to riparian, or riverbank, ecosystems as well as for the forest products industry. A native fungal pathogen, *Septoria musiva*, has begun moving west, attacking natural populations of black cottonwood in Pacific Northwest forests and intensively cultivated hybrid poplar in Ontario. And the infamous chestnut blight, a fungus accidentally introduced from Asia to North America in the late 1800s, wiped out billions of American chestnut trees.

Can biotech come to the rescue? Should it?

It's Complicated

Although there are many potential applications of biotechnology in forests, such as genetically engineering insects to suppress pest populations, we focused specifically on biotech trees that could resist pests and pathogens. Through genetic engineering, for example, researchers could insert genes, from a similar or unrelated species, that would help a tree tolerate or fight an insect or fungus.

It's tempting to assume that the buzz and enthusiasm for gene editing will guarantee quick, easy, and cheap solutions to these problems. But making a biotech tree will not be easy. Trees are large and long-lived, which means that the research required to test the durability and stability of an introduced trait will be expensive and take decades or longer. We also don't know nearly as much about the complex and enormous genomes of trees, compared with lab favorites such as fruit flies and the mustard plant, *Arabidopsis*.

In addition, because trees need to survive over time and adapt to a changing environment, it is essential to preserve and incorporate their existing genetic diversity into any "new" tree. Through evolutionary processes, tree populations already have many important adaptations to varied threats, and losing those could be disastrous. So even the fanciest biotech tree will ultimately depend on a thoughtful and deliberate breeding program to ensure its long-term survival. For these reasons, the committee called forth by the National Academies of Sciences, Engineering, and Medicine recommended increasing investment not just in biotechnology research but also in tree breeding, forest ecology, and population genetics.

Oversight Challenges

The committee found that the US Coordinated Framework for the Regulation of Biotechnology, which distributes federal oversight of biotech products among the Environmental Protection Agency, the US Department of Agriculture (USDA), and the Food and Drug Administration, is not fully prepared to consider the introduction of a biotech tree to improve forest health.

One blockage to approval is that regulators have always required the containment of pollen and seeds during biotech field trials to avoid the escape of genetic material. For example, the biotech chestnut was not allowed to flower so that transgenic pollen wouldn't blow across the landscape during field trials.[3] But if biotech trees are intended to spread their new traits, via seeds and pollen, and so introduce pest resistance across forests, then studies of wild reproduction will be necessary. These are rarely allowed until a biotech tree is fully deregulated.

Another shortcoming of the current framework is that some biotech trees may not require any special review at all. The USDA, for example, was asked to consider a loblolly pine that was genetically engineered for greater wood density. But because the USDA's regulatory authority stems from its oversight of plant pest risks, it decided that it did not have any regulatory authority over that biotech tree. Similar questions arise regarding organisms whose genes are edited using new tools such as CRISPR.

The committee noted that US regulations fail to promote a comprehensive consideration when it comes to forest health. Although the National Environmental Policy Act sometimes helps, some risks and many potential benefits are unlikely to be evaluated. This is the case for biotech trees as well as other tools to counteract pests and pathogens, such as tree breeding, pesticides, and site management practices.

How Do You Measure the Value of a Forest?

The report of the National Academies of Sciences, Engineering, and Medicine suggests an "ecosystem services" framework for considering the various ways that trees and forests provide value to humans. These ways range from extraction of forest products to the use of forests for recreation to the ecological services a forest provides, chiefly water purification, species protection, and carbon storage.

The committee also acknowledged that some ways of valuing a forest do not fit into the ecosystem services framework. For example, if forests are seen by some to have *intrinsic* value, then they have value in and of themselves, apart from the way humans value them and perhaps implying a kind of moral obligation to protect and respect them. Issues of "wildness" and "naturalness" also surfaced.

Wild Nature?

Paradoxically, a biotech tree could either increase or decrease wildness, depending on one's view. If wildness requires a lack of human intervention, then a biotech tree will reduce the wildness of a forest. But what if the introduction of a biotech tree were to prevent the eradication of an important tree species in the wild? There are no right or wrong answers to a question such as this, but it reminds us of the complexity of decisions to use technology to enhance nature.

This complexity points to a key recommendation in the report of the National Academies of Sciences, Engineering, and Medicine: to have dialogue among experts, stakeholders, and communities about how to value forests, to assess the risks and potential benefits of biotech, and to understand complex public responses to any potential interventions, including those involving biotechnology. These dialogic processes need to be respectful, deliberative, transparent, and inclusive. Such processes, like a 2018 stakeholder workshop on the biotech chestnut,[4] will not erase conflict or even guarantee consensus, but they can create an understanding that informs democratic decisions with expert knowledge and public values.

Notes

1. Krist, F. J., Ellenwood, J. R., Woods, M. E., McMahan, A. J., Cowardin, J. P., Ryerson, D. E., Sapio, F. J., Zweifler, M. O., & Romero, S. A. (2014, January). *2013–2027 national insect and disease forest risk assessment*. US Forest Service, United States Department of Agriculture. https://www.fs.fed.us/foresthealth/technology/pdfs /2012_RiskMap_Exec_summary.pdf.
2. Offutt, S. E., Delborne, J. A., DiFazio, S., & Ibáñez, I. (2019, January 8). *Forest health and biotechnology: Possibilities and considerations*. National Academies of Sciences, Engineering, and Medicine. https://doi

.org/10.17226/25221. The committee's purpose is set forth in Box S-1 of the report.

3. *Restoring the American chestnut.* (n.d.). SUNY College of Environmental Science and Forestry. https://www.esf.edu/chestnut/.

4. Delborne, J. A., Binder, A. R., Rivers, L., Barnes, J. C., Barnhill-Dilling, K., George, D., Kokotovich, A., & Sudweeks, J. (2018). *Biotechnology, the American chestnut tree, and public engagement workshop report.* North Carolina State University. https://research.ncsu.edu/ges/files/2018/10/Biotech-American-Chestnut-Public-Engagement-2018.pdf.

Chapter 14

How Scientists Make Plant-Based Foods Taste and Look More Like Meat

MARIANA LAMAS

IN 2019 BURGER KING SWEDEN released a plant-based burger, the Rebel Whopper, and the reaction was underwhelming. So, the company challenged its customers to taste the difference.

Burger King Sweden created a menu item for which customers had a 50–50 chance of getting a meat burger or a plant-based one. To find out which one they had eaten, they scanned the burger box with Burger King's app. The results: 44 percent guessed wrong—customers couldn't tell the difference.

Plant-based meats are products designed to imitate meat. While earlier products like tofu and seitan were meant to replace meat, newer products are trying to mimic its taste, texture, smell, and appearance. Plant-based burgers, ground meat, sausages, nuggets, and seafood are now in grocery stores and on restaurant menus. They aim to redefine our understanding of meat.

Achieving satisfactory mimicry is not an easy task. It took Beyond Meat more than six years to develop the Beyond Burger. And since its release in 2015, it has been through three reformulations. The science behind building the perfect plant-based meat is full of trial and error—and involves a multidisciplinary team.

The Maillard Reaction

Appearance, texture, and flavor are the three main challenges that food scientists face when developing a convincing plant-based meat. These are what give meat its characteristics and essence. When meat cooks, its texture changes. The temperature of the pan or the grill affects protein structures.[1] As proteins begin to break down, coagulate, and contract, the meat tenderizes and holds its shape. What's known as the Maillard reaction is responsible for that distinctive "meaty" aroma and savory flavor. Understanding it helps food research and development teams replicate it in plant-based meat products.

Ingredients also influence appearance, texture, and flavor. Soy, wheat, pea, and fava proteins, as well as starches, flours, hydrocolloids (nondigestible carbohydrates used as thickeners, stabilizers, and emulsifiers or as water-retaining and gel-forming agents), and oils, can make a plant-based meat more or less similar to the animal meat it is trying to emulate.

Finally, the processing method influences the product's final characteristics. "High-moisture extrusion" and "shear cell" technologies are two of the most common processes used to transform vegetable protein into a layered fibrous structure that closely matches the appearance and texture of meat. High-moisture extrusion is the most used technique and provides a meat-like bite,[2] but shear-cell processing is more energy-efficient and has a smaller carbon footprint.[3]

Color and Texture

Food scientists are now able to simulate meat color before, during, and after cooking. Beet extract, pomegranate powder, and soy leghemoglobin have been used to mimic the red color of fresh or rare beef.

Animal protein texture, though, is difficult to copy with plant-based ingredients because plants do not have muscle tissue. Muscles are elastic and flexible, while plant cells are rigid and unbending. Plants do not have the bite and chewiness of meat, which is why veggie burgers can often feel crumbly and mushy.

A key ingredient in any plant-based meat is the plant protein. In addition to being fundamental to the structure,[4] it is important for product identity and differentiation as well. A formulation can use one type of protein or a blend of different types. The most common of meat mimics, soy protein is still the plant protein that delivers the most meat-like taste and texture. Since it has been used for decades now, a lot of research has been done, and its texturization process has been improved.

Pea protein, made popular by Beyond Meat, is the fastest-growing segment in the plant-based market, however, because of its complete amino acid profile. There are nine amino acids that are essential in our diets. Animal-based foods have all of them and are considered complete proteins. Most plant foods are incomplete proteins, meaning certain amino acids are missing, but pea protein contains all nine. Pea protein also lacks allergens. Rice, fava, chickpea, lentil, and mung bean proteins have also generated a lot of interest among food scientists, and more products incorporating these plants are expected to come to market in the future.

Creating Flavor

Companies don't have to disclose flavoring ingredients—only whether they're natural or artificial—so it's hard to know what exactly gives plant-based burgers that meat-like flavor.

Fat is a major player in flavor and mouth feel. It provides mouth-coating richness and juiciness and is responsible for flavor release. It activates certain areas of the brain that are responsible for processing taste, aroma, and reward mechanisms. The industry standard has been to use coconut oil to replace animal fat. However, coconut oil melts at much a lower temperature than animal fat. In the mouth this translates to bites that start off rich and juicy but wear off quickly. Some plant-based meats use a combination of plant-based oils, such as canola and sunflower oils, to increase the melting temperature and extend the juiciness. New replacements for animal fats using sunflower oil and water emulsions and cultivated animal fats (fat cells grown in laboratories) are being developed to solve this problem. But clearly, not all of these would suit a vegetarian or vegan diet.

A plant-based meat formulation can work on paper, have the recommended number of ingredients, and hit the nutritional targets to match meat, but it might not taste good or have the right texture or bite. For example, potato protein creates great texture, but it is bitter. Food scientists must find a balance between the protein content, texture, and flavor.

The Future of Formulated Food

Food scientists have only scratched the surface when it comes to unlocking the potential of plant-based meats. There is still a lot to explore and improve. The current commercially available plant protein ingredients come from 2 percent of about 1,505 plant protein species used for food supply.[5]

There is ongoing research exploring crop optimization through breeding or engineering to increase protein content to support the further development and improvement of plant protein isolates and ultimately plant-based meats. Processing-method technologies are still being developed, and we are seeing new technologies such as 3D printing and cultured meat being adopted and refined. Expect to see plant-based meat products increase and whole cuts, like steaks, to be commercially available soon.

Notes

1. Tornberg, E. (2005). Effects of heat on meat proteins—implications on structure and quality of meat products. *Meat Science*, *70*(3), 493–508. https://doi.org/10.1016/j.meatsci.2004.11.021.
2. Osen, R., & Schweiggert-Weisz, U. (2016). High-moisture extrusion: Meat analogues. *Reference Module in Food Science*. https://doi.org /10.1016/b978-0-08-100596-5.03099-7.
3. Cornet, S. H., Snel, S. J., Schreuders, F. K., van der Sman, R. G., Beyrer, M., & van der Goot, A. J. (2021). Thermo-mechanical processing of plant proteins using shear cell and high-moisture extrusion cooking. *Critical*

    Reviews in Food Science and Nutrition, 1–18. https://doi.org/10.1080
    /10408398.2020.1864618.

4. He, J., Evans, N. M., Liu, H., & Shao, S. (2020). A review of research on
    plant-based meat alternatives: Driving forces, history, manufacturing,
    and consumer attitudes. *Comprehensive Reviews in Food Science and
    Food Safety, 19*(5), 2639–2656. https://doi.org/10.1111/1541-4337
    .12610.

5. Green, M. (2020, July 14). *Shift20: Industry is "only scratching the
    surface" of plant-based proteins.* Food Ingredients First. https://www
    .foodingredientsfirst.com/news/shift20-industry-is-only-scratching
    -the-surface-of-plant-based-proteins.html.

Chapter 15

Genetically Modified Mosquitoes May Be the Best Weapon for Curbing Disease Transmission

JASON RASGON

MOSQUITOES ARE SOME OF THE DEADLIEST creatures on the planet. They carry viruses, bacteria, and parasites, which they transmit through bites, infecting some 700 million people and killing more than a million each year.

With international travel, migration, and climate change, these infections are no longer confined to tropical and subtropical developing countries. Pathogens such as West Nile virus and Zika virus have caused significant outbreaks in the United States and its territories that are likely to continue, with new invasive pathogens being discovered all the time. Currently, control of these diseases is mostly limited to broad-spectrum insecticide sprays, which can harm both humans and non-target animals and insects. What if there was a way to control these devastating diseases without the environmental problems of widespread insecticide use?

Genetically modifying mosquitoes to prevent disease may sound like science fiction, but the technology has advanced in recent years to the point where this is no longer a scenario fit only for late-night movies. In fact, it's not even a new idea; scientists were talking about modifying insect populations to control diseases as early as the 1940s.[1] Today, genetically modified (GM) mosquitoes, developed through decades of research in university laboratories, are being used to combat mosquito-borne pathogens—including viruses such as dengue and Zika—in many locations around the globe, including the United States. Progress is also being made to use GM mosquitoes to combat malaria, the most devastating mosquito-borne disease, although field releases for malaria control have not yet taken place.

I have been working on GM mosquitoes, both as a lab tool and to combat disease, for almost 25 years. During that time, I have personally witnessed the technology go from theo-retical in the lab to applied in the field. I've seen older tech-niques that were inefficient, random, and slow pave the way for new methods such as CRISPR, which enables efficient, rapid, and precise editing of mosquito genomes, and like ReMOT Control, which eliminates the requirement of injecting materials into mosquito embryos, thus making the generation

of modified mosquitoes much easier. These new technologies make GM mosquitoes for disease control not a question of "if" but rather a question of "where" and "when."

Don't worry; these genetic changes only affect the mosquitoes—they are not transmitted to people when the mosquitoes bite them.

Ways to Use Genetically Modified Mosquitoes

There are two methods currently used to control mosquito-borne diseases with GM mosquitoes. The first is *population replacement* in which a mosquito population that is biologically able to transmit pathogens is replaced by one that is unable to transmit pathogens. This approach generally relies on a concept known as gene drive to spread the anti–pathogen genes. In gene drive, a genetic trait—a gene or group of genes—relies on a quirk of inheritance to spread to more than

A worker sprays anti-mosquito fog in an attempt to control dengue fever in a neighborhood in Jakarta, Indonesia. Highly populated areas in the country are often hit with severe outbreaks of the mosquito-borne disease, especially during the annual rainy season, due to poor health services and unsanitary living conditions. *AP Photo / Achmad Ibrahim*

half of a mosquito's offspring, thereby boosting the frequency of the trait in the population. The second method is *population suppression*. This strategy reduces mosquito populations so that there are fewer mosquitoes to pass on the pathogen.

While the concept of gene drive in mosquitoes is many decades old, the gene-editing technique CRISPR has finally made it possible to engineer it easily in the laboratory. CRISPR-based gene drives have not yet been deployed in nature, however, mostly because they are still a new technology that lacks a firm international regulatory framework but also because of problems related to the evolution of resistance in mosquito populations that will stop the gene from spreading.

It may not be immediately obvious, but the gene in a gene drive need not be a gene at all—it can be a microbe. All organisms exist not just with their own genomes but also with the genomes of all their associated microbes: the hologenome, as it is known. The spread of a microbial genome through a population by inheritance can also be thought of as gene drive. By this definition, the first gene drive that has been deployed in mosquito populations for disease control is a bacterial symbiont known as *Wolbachia*. This bacterium infects up to 70 percent of all known insect species, where it hijacks insect reproduction to spread itself through the population.

Thus, *Wolbachia* itself (with its genome of approximately 1,500 genes) acts as the genetic trait that is driven into the population. When *Wolbachia* is transferred into a previously uninfected mosquito, it often makes the mosquito more resistant to infection with pathogens that can cause disease in humans, such as multiple viruses (including dengue and Zika viruses) and malaria parasites.

A Bacterium That Fights Disease

Over the past decade, researchers have taken *Wolbachia* present in fruit flies and transferred it to mosquitoes that transmit dengue virus. The modified insects were then released in a dozen countries to control the disease. Although marketed as a "non-GM strategy," artificially infecting mosquitoes with *Wolbachia* clearly falls under the GM umbrella, as over 1,500 genes (the entire bacterial genome) have been transferred from the original fruit fly host to the mosquitoes.

Preliminary results for suppressing dengue from these releases in Australia were promising, leading to a randomized control trial of the technology in Indonesia, which demonstrated a 77 percent reduction in dengue virus incidence in release sites compared with control sites.[2] However, similar suppression in other release areas with higher disease risk, such as South America and Asia, still needs to be determined, particularly as some studies have demonstrated that *Wolbachia* can sometimes increase pathogen infection in mosquitoes rather than suppress it.[3]

GM Mosquitoes That Eliminate Mosquitoes

The best current example of population suppression is the release of GM sterile mosquitoes. This is a modern spin on the decades-old sterile insect technique, where sterile male insects are released into natural populations to mate with the wild females, reducing the mosquito population. But rather than crudely sterilizing mosquitoes with radiation or chemicals, clever genetic engineering is now used to sterilize them instead. The company Oxitec has engineered mosquitoes with a gene that is lethal to females but not to males, which do not bite or transmit disease. Thousands of these transgenic males are released into nature, where they mate with the wild females in the population. The genetic modification is inherited by the offspring of these matings; female offspring die, while male offspring, which carry the gene, survive and continue passing the trait to future generations. With fewer and fewer females, the mosquito population is drastically suppressed. Oxitec has conducted releases in the Grand Caymans, Malaysia, Brazil, and Florida.

There has been some opposition to these sterile mosquito releases, particularly in Florida. For example, in 2016, an Oxitec trial in the Florida Keys was met with some local resistance. However, unlike gene drive strategies, releasing sterile mosquitoes (GM or not) has about the smallest environmental footprint and highest safety of any disease-control strategy; certainly it is safer than broad-spectrum insecticide sprays. It is highly targeted and thus, if it works, will only result in eliminating the target mosquito species, which in this case (*Aedes aegypti*) is a highly invasive and non-native mosquito in Florida. In 2021 Oxitec conducted the first field releases of its sterile mosquitoes in Florida.[4]

In addition to gene drive, *Wolbachia* bacteria have also been used for population suppression. Males infected with the bacteria are released into a mosquito population that is either not infected or infected with a different *Wolbachia* strain, which leads to "incompatible," or sterile, matings. This strategy, again, has a long history and was first used to suppress mosquito populations in the 1960s before people even knew that *Wolbachia* was causing certain populations of mosquitoes to be sterile when mated with one another. In current times, *Wolbachia*-sterilized males have been released in multiple countries including Australia and the United States, in California and Florida, to control dengue virus.[5]

In an increasingly interconnected world, and with the added problems of global climate change, pathogens are not likely to stay confined to the developing world but will become an increasing issue for the United States as well. With the evolution of insecticide resistance in mosquitoes being a certainty, GM technology has the potential to reduce the burden of mosquito-borne diseases around the globe, without the environmental and health risks associated with harmful pesticide use.

Don't be afraid if it still sounds like science fiction; it may just save your life.

Notes

1. Macias, V., Ohm, J., & Rasgon, J. (2017). Gene drive for mosquito control: Where did it come from and where are we headed? *International Journal of Environmental Research and Public Health, 14*(9), 1006. https://doi.org/10.3390/ijerph14091006.
2. Utarini, A., Indriani, C., Ahmad, R. A., Tantowijoyo, W., Arguni, E., Ansari, M. R., Supriyati, E., Wardana, D. S., Meitika, Y., Ernesia, I., Nurhayati, I., Prabowo, E., Andari, B., Green, B. R., Hodgson, L., Cutcher, Z., Rancès, E., Ryan, P. A., O'Neill, S. L., . . . Simmons, C. P. (2021). Efficacy of Wolbachia-infected mosquito deployments for the control of dengue. *New England Journal of Medicine, 384*(23), 2177–2186. https://doi.org/10.1056/nejmoa2030243.
3. Dodson, B. L., Hughes, G. L., Paul, O., Matacchiero, A. C., Kramer, L. D., & Rasgon, J. L. (2014). *Wolbachia* enhances West Nile virus (WNV)

infection in the mosquito *Culex tarsalis*. *PLoS Neglected Tropical Diseases, 8*(7). https://doi.org/10.1371/journal.pntd.0002965.

4. Fensom, M. (2021, December 22). *Oxitec successfully concludes mosquito releases in landmark Florida Keys pilot program*. Oxitec. https://www.oxitec.com/en/news/oxitec-successfully-concludes -mosquito-releases-in-landmark-florida-keys-pilot-program.

5. Mains, J. W., Brelsfoard, C. L., Rose, R. I., & Dobson, S. L. (2016). Female adult *Aedes albopictus* suppression by *Wolbachia*-infected male mosquitoes. *Scientific Reports, 6*(1). https://doi.org/10.1038 /srep33846.

Chapter 16

How Engineered Bacteria Could Clean Up Oil Sands Pollution and Mining Waste

VIKRAMADITYA G. YADAV

RAMPANT INDUSTRIALIZATION HAS CAUSED our planet to warm at an unprecedented rate. Glaciers are melting away, and sea levels are rising. Droughts last longer and are more devastating. Forest fires are more intense. Extreme, once-in-a-generation weather events—such as Category 5 hurricanes—seem to be occurring on an annual basis. The environment is indeed in grave danger, and urgent action is desperately needed. But there is genuine optimism that solutions to some of the largest environmental challenges may finally be at hand.

Take, for example, the decades-long problem of oil sands' tailings ponds in Canada, the third-largest reserve of crude oil in the world. The recovery of this oil consumes nearly threefold its volume in water and leaves behind a slurry of water, solids, and organic contaminants as waste. Oil sands operations are into their seventh decade, and more than a trillion liters of wastewater now resides in tailings ponds.

But a rapidly growing collective of engineers, scientists, activists, and entrepreneurs are delivering some of the biggest gains in environmental remediation in recent decades by blurring the lines between physical, biological, and digital sciences. We call ourselves synthetic biologists. I have contributed extensively to research, education, commercialization, and regulation of synthetic biology, including as the founder of Metabolik Technologies, an environmental biotechnology venture, which commercialized a first-of-its-kind, low-energy, low-cost, and sustainable solution to decontaminate oil sands' tailings ponds.

A Quick Guide to Synthetic Biology

The underlying premise of synthetic biology is as simple as it is elegant: Nature assembles, dismantles, and recycles molecules in the cleanest and most efficient manner imaginable. The unique instructions required to achieve these tasks are found in DNA. Synthetic biologists investigate natural systems to understand their remarkable processes and then use lab-synthesized DNA to reprogram them to perform new tasks or existing tasks more efficiently.

Synthetic biology has been used to improve enzymes, cells, and populations of cells for diverse applications such as sensing,[1] breaking down hydrocarbons and other "forever chemicals" such as per- and polyfluoroalkyl substances in soil and water,[2] and sequestering carbon dioxide and methane.[3]

Once Fiction, Now a Real Solution

Importantly, many of the protagonists and influencers of synthetic biology are millennials and zoomers (Gen Z) who were raised on a steady diet of Saturday morning cartoons. Genetically engineered bacteria that mopped up oil spills were a staple in *Captain Planet and the Planeteers*, an animated environmental superhero series that launched in 1990. Whereas two decades ago such a concept was confined to science fiction, it is now a reality, owing to advances in molecular biology such as CRISPR genome editing and the advent of fully automated genomic foundries—robotic systems that conduct thousands of experiments a day—for accelerated design-build-test-learn cycles.[4]

Crucially, the successes of synthetic biology in the sphere of environmental remediation have not been one-off demonstrations in academic laboratories. They have been proven in the field at sizable scales, and they have taken large bites out of some of the greatest environmental challenges in the world.

Translating Innovations to the Field

Tailings ponds contain organic compounds such as naphthenic acid fraction compounds and polyaromatic hydrocarbons that are harmful to aquatic life and human health,[5] and they are notoriously difficult to eliminate from water.[6] They are also teeming with microbial life. These microbes do not merely survive but thrive in the contaminated water. They sense, ingest, and metabolize the toxic compounds in the water, albeit at very slow rates.

My group at the University of British Columbia and our colleagues at Allonnia isolated and studied the genomics of these unique creatures and, in collaboration with Ginkgo Bioworks, have worked on increasing their appetite for and metabolism of the toxic compounds. After validating the performance of the microorganisms in the field, the UBC–Allonnia team designed some of the largest treatment systems of their kind to achieve the rates and scales needed to remediate the contaminated water within the timeline prescribed in Alberta's Tailings Management Framework.[7]

We have tested our treatment system in demonstration systems that mimic the complexity of tailings ponds in order to fine-tune the microorganisms and the reactors and assess the risks. Some of these risks include the inefficacy or higher-than-expected costs of the technology, the potential damage the microbes may cause to the ecosystem, and the resistance that regulators and shareholders may have to deploying engineered microorganisms in the environment.

This small team of synthetic biologists succeeded thanks to the ingenuity of the approach and new models of collaboration. The team also involved oil sands operators, engineering design firms, contract manufacturing companies, and regulatory experts who were able to leverage each partner's strengths to reduce the time, expense, and uncertainty of developing a practical solution.

The Fun Has Only Just Begun

Synthetic biologists are only getting started and have now set their sights on a number of similarly large problems. One, in particular, has significant implications for our electric future.

Widespread adoption of electric vehicles could reduce carbon emissions from the transportation sector by nearly 50 percent. Unfortunately, mining the metals used in electric vehicles damages the environment. The manufacture of a single electric vehicle generates 250,000 kilograms of mining waste and 150,000 liters of an extremely toxic liquid called acid rock drainage, a major threat to the environment owing to its potentially devastating effect on aquatic habitats.

Mining is wasteful and unsustainable, and the industry is in desperate need of effective solutions to treat its large bodies of waste. My startup company Tersa Earth is developing microbial solutions to do away with tailings ponds entirely. If we and others like us are successful, we will help to eliminate waste, deliver decarbonization, preserve biodiversity, generate employment, and achieve equitable social development.

Notes

1. Del Valle, I., Fulk, E. M., Kalvapalle, P., Silberg, J. J., Masiello, C. A., & Stadler, L. B. (2021). Translating new synthetic biology advances for biosensing into the earth and environmental sciences. *Frontiers in Microbiology, 11.* https://doi.org/10.3389/fmicb.2020.618373.

2. Genovese, M., Denaro, R., Cappello, S., Di Marco, G., La Spada, G., Giuliano, L., Genovese, L., & Yakimov, M. M. (2008). Bioremediation of benzene, toluene, ethylbenzene, xylenes-contaminated soil: A biopile pilot experiment. *Journal of Applied Microbiology, 105*(5), 1694–1702. https://doi.org/10.1111/j.1365-2672.2008.03897.x.

3. DeLisi, C., Patrinos, A., MacCracken, M., Drell, D., Annas, G., Arkin, A., Church, G., Cook-Deegan, R., Jacoby, H., Lidstrom, M., Melillo, J., Milo, R., Paustian, K., Reilly, J., Roberts, R. J., Segrè, D., Solomon, S., Woolf, D., Wullschleger, S. D., & Yang, X. (2020). The role of synthetic biology in atmospheric greenhouse gas reduction: Prospects and challenges. *BioDesign Research, 2020,* 1–8. https://doi.org/10.34133/2020/1016207.

4. Chao, R., Mishra, S., Si, T., & Zhao, H. (2017). Engineering biological systems using automated biofoundries. *Metabolic Engineering, 42,* 98–108. https://doi.org/10.1016/j.ymben.2017.06.003.

5. Bauer, A. E., Hewitt, L. M., Parrott, J. L., Bartlett, A. J., Gillis, P. L., Deeth, L. E., Rudy, M. D., Vanderveen, R., Brown, L., Campbell, S. D., Rodrigues, M. R., Farwell, A. J., Dixon, D. G., & Frank, R. A. (2019). The toxicity of organic fractions from aged oil sands process-affected water to aquatic species. *Science of The Total Environment*, *669*, 702–710. https://doi.org/10.1016/j.scitotenv.2019.03.107.

6. Quinlan, P. J., & Tam, K. C. (2015). Water treatment technologies for the remediation of naphthenic acids in oil sands process-affected water. *Chemical Engineering Journal*, *279*, 696–714. https://doi.org/10.1016/j.cej.2015.05.062.

7. Chegounian, P., Zerriffi, H., & Yadav, V. G. (2020). Engineering microbes for remediation of oil sands tailings. *Trends in Biotechnology*, *38*(11), 1192–1196.

Part III

Powerful Tools for Medicine and Health

One of the fastest-growing and most visible areas of bio-technological innovation is medicine, with new technologies opening up novel possibilities for treatment on a regular basis. Growing human organoids in a petri dish, 3D-printing scaffolds to fix nerve damage, and searching for longevity genes are no longer the subjects of science fiction; instead they are activities of lab research and are making their way to biotech startups. Mice are commonly used model systems in scientific research, but just because something works in mice doesn't mean it will act the same in humans. In some labs, humanized pigs have been created as more realistic model systems for research on human stem cells and for growing organs that can be donated to humans (xenotransplantation). The biotech industry draws on decades of research in corporate and university labs to develop and harness the latest biotechnologies to improve human health.

The chapters of part III discuss the potential of today's advances in biotechnology to address some of the most pressing problems in medicine and also describe trailblazing experiments that show future directions for the biosciences. The power and shortcomings of CRISPR are mentioned in multiple chapters, and contributors explain how model organisms are used for the long-term goal of improving human patient outcomes. Significant financial investments are required for moving these exciting developments through clinical trials (testing for safety and efficacy) to clinical applications.

Based on the past, it is safe to predict that many of the advances discussed here will not make it to the clinic. The investors in these biotechnologies obviously want to recoup their money, which can result in exorbitant prices for new treatments that treat diseases affecting a small number of patients. Similar treatments with a small market may never make it to the clinic. The ethical questions brought up in part III will be examined in more detail in part IV.

Chapter 17

New Gene Therapies May Soon Treat Dozens of Rare Diseases, but Million-Dollar Price Tags Will Put Them out of Reach for Many

KEVIN DOXZEN

ZOLGENSMA—WHICH TREATS spinal muscular atrophy, a rare genetic disease that damages nerve cells and leads to muscle decay—currently holds the world record for the most expensive drug to buy. A one-time treatment of the lifesaving drug for a young child costs US$2.1 million. While Zolgensma's exorbitant price is an outlier today, by the end of the decade there'll be dozens of cell and gene therapies that cost hundreds of thousands to millions of dollars for a single dose. The Food and Drug Administration predicts that, by 2025, it will be approving 10 to 20 cell and gene therapies every year.

I'm a biotechnology and policy expert focused on improving access to cell and gene therapies. While these forthcoming treatments have the potential to save many lives and ease much suffering, health care systems around the world aren't equipped to handle them. Creative new payment systems will be necessary to ensure that everyone has equal access to these therapies.

The Rise of Gene Therapies

Thousands of diseases are the result of DNA errors, which prevent cells from functioning normally.[1] Currently, only 5 percent of the roughly 7,000 rare diseases have an FDA-approved drug, leaving thousands of conditions without a pharmaceutical treatment. But over the past few years, genetic engineering technology has made impressive strides toward the ultimate goal of curing disease by changing a cell's genetic instructions. By directly correcting disease-causing mutations or altering a cell's DNA to give the cell new tools to fight disease, gene therapy offers a powerful new approach to medicine. The resulting gene therapies will be able to treat many diseases at the DNA level in a single dose.

In 2022 there were more than 2,000 gene therapies in development around the world. A large fraction of this research focuses on rare genetic diseases, which affect 400 million people worldwide. We may soon see cures for rare diseases like sickle cell disease,[2] muscular dystrophy,[3] and progeria, a rare and progressive genetic disorder that causes children to age rapidly. Farther into the future, gene therapies may treat more common conditions, like heart disease and chronic pain.

Sky-High Price Tags

The problem is that these therapies will carry shocking price tags.

Gene therapies are the result of years of research and development totaling hundreds of millions to billions of dollars. Sophisticated manufacturing facilities, highly trained personnel, and complex biological materials set gene therapies apart from other drugs. To recoup costs, especially for drugs with a small number of potential patients, pharmaceutical companies have to set higher prices.

The toll of high prices on health care systems will not be trivial. Consider a gene therapy cure for sickle cell disease, which is expected to be available in the next few years. The estimated price of this treatment is $1.85 million per patient. As a result, economists predict that it could cost a single state's Medicare program almost $30 million per year,[4] even when assuming that only 7 percent of the eligible population receive the treatment. And that's just one drug. Introducing dozens of similar therapies into the market would strain health care systems and create difficult financial decisions for private insurers.[5]

Lowering Costs, Finding New Ways to Pay

One solution for improving patient access to gene therapies would be simply to demand that drugmakers charge less money for their products, a tactic taken in Germany. But this approach comes with a lot of challenges and may mean that companies simply refuse to offer the treatment in certain places.

I think a more balanced and sustainable approach is twofold. In the short term, it'll be important to develop new payment methods that entice insurance companies to cover high-cost therapies and distribute risks across patients, insurance companies, and drugmakers. In the long run, improved gene therapy technology will inevitably help lower costs.

For innovative payment models, one tested approach is tying coverage to patient health outcomes. Since these therapies are still experimental and relatively new, there isn't much data to help insurers make the risky decision to cover them. If an insurance company is paying $1 million for a therapy, it had better work. In an outcomes-based model, insurers will either pay for some of the therapy up front and the rest only if the patient improves or, alternatively, cover the entire cost up front and receive a reimbursement if the patient doesn't get better. These options help insurers share financial risk with the drug developers.

Another model is known as the "Netflix model" and would act as a subscription-based service. In this model, a state Medicaid program would pay a pharmaceutical company a flat fee for access to unlimited treatments. This would allow a state to provide the treatment to residents who qualify and would help governments balance their budgets while giving drugmakers money up front. This model has worked well for improving access to hepatitis C drugs in Louisiana.

On the cost front, the key to improving access will be investing in new technologies that simplify medical procedures. For example, the costly sickle cell gene therapies currently in clinical trials require a series of expensive steps, including a stem cell transplant. The Bill & Melinda Gates Foundation, the National Institutes of Health, and Novartis are partnering to develop an alternative approach that would involve a simple injection of gene therapy molecules. The goal of their collaboration is to help bring an affordable sickle cell treatment to patients in Africa and other low-resource settings.

Improving access to gene therapies requires collaboration and compromise across governments, nonprofits, pharmaceutical companies, and insurers. Taking proactive steps now to develop innovative payment models and invest in new technologies will help ensure that health care systems are ready to deliver on the promise of gene therapies.

Notes

1. Prakash, V., Moore, M., & Yáñez-Muñoz, R. J. (2016). Current progress in therapeutic gene editing for monogenic diseases. *Molecular Therapy, 24*(3), 465–474. https://doi.org/10.1038/mt.2016.5.
2. Frangoul, H., Altshuler, D., Cappellini, M. D., Chen, Y.-S., Domm, J., Eustace, B. K., Foell, J., de la Fuente, J., Grupp, S., Handgretinger, R., Ho, T. W., Kattamis, A., Kernytsky, A., Lekstrom-Himes, J., Li, A. M., Locatelli, F., Mapara, M. Y., de Montalembert, M., Rondelli, D., . . . Corbacioglu, S. (2021). CRISPR-Cas9 gene editing for sickle cell disease and β-thalassemia. *New England Journal of Medicine, 384*(3), 252–260. https://doi.org/10.1056/nejmoa2031054.
3. Olson, E. N. (2021). Toward the correction of muscular dystrophy by gene editing. *Proceedings of the National Academy of Sciences, 118*(22). https://doi.org/10.1073/pnas.2004840117.
4. DeMartino, P., Haag, M. B., Hersh, A. R., Caughey, A. B., & Roth, J. A. (2021). A budget impact analysis of gene therapy for sickle cell disease. *JAMA Pediatrics, 175*(6), 617–623. https://doi.org/10.1001/jamapediatrics.2020.7140.
5. Quinn, C., Young, C., Thomas, J., & Trusheim, M. (2019). Estimating the clinical pipeline of cell and gene therapies and their potential economic impact on the US healthcare system. *Value in Health, 22*(6), 621–626. https://doi.org/10.1016/j.jval.2019.03.014.

Chapter 18

Engineered Viruses Can Fight the Rise of Antibiotic-Resistant Bacteria

KEVIN DOXZEN

AS THE WORLD FOCUSED ON FIGHTING the SARS-CoV-2 virus causing the COVID-19 pandemic, another group of dangerous pathogens loomed in the background. The threat of antibiotic-resistant bacteria has been growing for years and appears to be getting worse. If COVID-19 taught us one thing, it's that governments should be prepared for more global public health crises, and that includes finding new ways to combat rogue bacteria that are becoming resistant to commonly used drugs.

In contrast to the pandemic, a virus may be the hero of the next epidemic rather than the villain. Scientists have shown that viruses could be great weapons against bacteria that are resistant to antibiotics.

I am a biotechnology and policy expert focused on understanding how personal genetic and biological information can improve human health. Every person interacts intimately with a unique assortment of viruses and bacteria, and by deciphering these complex relationships, we can better treat infectious diseases caused by antibiotic-resistant bacteria.

Replacing Antibiotics with Bacteriophages

Since the discovery of penicillin in 1928, antibiotics have changed modern medicine. These small molecules fight off bacterial infections by killing or inhibiting the growth of bacteria. The mid-20th century was called the golden age for antibiotics, a time when scientists were discovering dozens of new molecules for many diseases.

This high was soon followed by a devastating low.[1] Researchers saw that many bacteria were evolving resistance to antibiotics. Bacteria in our bodies were learning to evade medicine by mutating and evolving to the point where antibiotics no longer worked.

As an alternative to antibiotics, some researchers are turning to a natural enemy of bacteria: bacteriophages. Bacteriophages are viruses that infect bacteria. They outnumber bacteria 10 to 1 and are considered to be the most abundant organisms on the planet.[2] Bacteriophages, or simply phages, survive by infecting a bacterium, replicating, and bursting out from their host, which destroys the bacterium.

Harnessing the power of phages to fight bacteria isn't a new idea. The first recorded use of so-called phage therapy was over a century ago. In 1919 French microbiologist Félix d'Hérelle used a cocktail of phages to treat children suffering from severe dysentery. He is credited with co-discovering phages,[3] and he pioneered the idea of using bacteria's natural enemies in medicine. He would go on to stop outbreaks of cholera in India and plague in Egypt.[4]

Phage therapy is not a standard treatment you can find in your local hospital today. But excitement about phages has grown over the past few years.[5] In particular, scientists are using new knowledge about the complex relationship between phages and bacteria to improve phage therapy. By engineering phages to better target and destroy bacteria, scientists hope to overcome antibiotic resistance.

Engineering Phages

Even if you are not a biologist, you may have heard of one type of bacterial immune system: CRISPR, which stands for clustered regularly interspaced short palindromic repeats.

Bacteriophages, illustrated here on the surface of bacteria, are viruses that infect and destroy bacteria. *Christoph Burgstedt / Science Photo Library, Getty Images*

This immune system helps bacteria store genetic information from viral infections. The bacteria then use that information to fight off future invaders, much as our own immune system can recognize a particular pathogen to fight off infection.

CRISPR proteins in bacteria locate and cut specific sequences of DNA or RNA found in viruses. Such precise cutting also makes CRISPR proteins efficient tools for editing the genomes of various organisms. This is why the development of CRISPR genome–editing technology won the Nobel Prize in Chemistry in 2020. Now scientists are hoping to use their knowledge of CRISPR systems to engineer phages to destroy dangerous bacteria.

When the engineered phage locates a specific bacterium, the phage injects CRISPR proteins inside the bacterium, cutting and destroying its DNA. Scientists have found a way to turn defense into offense. The proteins normally involved in protecting against viruses are repurposed to target and destroy the bacterium's own DNA. Scientists can specifically target the DNA that makes the bacterium resistant to antibiotics, making this type of phage therapy extremely effective.

The bacterium *Clostridioides difficile* is an antibiotic-resistant strain of bacteria that kills 29,000 people in the United States every year. In one demonstration of the CRISPR-based phage technique, researchers engineered phages to inject a molecule that directs the bacteria's own

CRISPR proteins to chew up the bacteria's DNA like a paper shredder.

CRISPR isn't the only bacterial immune system. Scientists are discovering more by using creative microbiological experiments and advanced computational tools. They have already found tens of thousands of new microbes,[6] along with dozens of new immune systems.[7] Scientists hope to find more tools that could help them engineer phages to target a wider range of bacteria.

Beyond the Science

Science is only half of the solution when it comes to fighting these microbes. Commercialization and regulation are important to ensure that this technology is in society's tool kit for fending off a worldwide spread of antibiotic-resistant bacteria.

Multiple companies are engineering phages, or identifying naturally occurring phages, to destroy specific harmful bacteria. Companies like Felix Biotechnology, Adaptive Phage Therapeutics, and Cytophage are producing specialized bacteria-killing phages to replace antibiotics in human health care and agriculture. BiomX aims to treat infections common in chronic diseases like cystic fibrosis and inflammatory bowel disease by using both natural and engineered phage cocktails. Thinking globally, the company PhagePro is using phages to treat cholera; these deadly bacteria primarily affect people in Africa and Asia.

Alongside the commercialization of phage therapy, policies that facilitate safe testing and regulation of the technology are vital. To avoid replicating America's poor COVID-19 response, I believe the world must invest in, engineer, and then test phage therapies. Proactive planning will help us combat whatever antibiotic-resistant bacteria might evolve and spread.

Notes

1. Hutchings, M. I., Truman, A. W., & Wilkinson, B. (2019). Antibiotics: Past, present and future. *Current Opinion in Microbiology, 51*, 72–80. https://doi.org/10.1016/j.mib.2019.10.008.

2. Stern, A., & Sorek, R. (2010). The phage-host arms race: Shaping the evolution of microbes. *BioEssays, 33*(1), 43–51. https://doi.org/10.1002/bies.201000071.

3. Fruciano, D. E., & Bourne, S. (2007). Phage as an antimicrobial agent: d'Herelle's heretical theories and their role in the decline of phage prophylaxis in the West. *Canadian Journal of Infectious Diseases and Medical Microbiology, 18*(1), 19–26. https://doi.org/10.1155/2007/976850.

4. Keen, E. C. (2012). Phage therapy: Concept to cure. *Frontiers in Microbiology, 3*. https://doi.org/10.3389/fmicb.2012.00238.

5. Schooley, R. T., & Strathdee, S. (2020). Treat phage like living antibiotics. *Nature Microbiology, 5*(3), 391–392. https://doi.org/10.1038/s41564-019-0666-4.

6. Nayfach, S., Roux, S., Seshadri, R., Udwary, D., Varghese, N., Schulz, F., Wu, D., Paez-Espino, D., Chen, I.-M., Huntemann, M., Palaniappan, K., Ladau, J., Mukherjee, S., Reddy, T. B., Nielsen, T., Kirton, E., Faria, J. P., Edirisinghe, J. N., Henry, C. S., . . . Eloe-Fadrosh, E. A. (2020). A genomic catalog of Earth's microbiomes. *Nature Biotechnology, 39*(4), 499–509. https://doi.org/10.1038/s41587-020-0718-6.

7. Doron, S., Melamed, S., Ofir, G., Leavitt, A., Lopatina, A., Keren, M., Amitai, G., & Sorek, R. (2018). Systematic discovery of antiphage defense systems in the microbial pangenome. *Science, 359*(6379). https://doi.org/10.1126/science.aar4120.

Chapter 19

Genetic Engineering Transformed Stem Cells into Working Mini-livers That Extended the Life of Mice with Liver Disease

MO EBRAHIMKHANI

IMAGINE IF RESEARCHERS could program stem cells, which have the potential to grow into all cell types in the body, so that they could generate an entire human organ. This would allow them to manufacture tissues for testing drugs and would reduce the demand for transplant organs by having new ones grown directly from a patient's cells.

I'm a researcher working in this new field—called synthetic biology—focused on creating new biological parts and redesigning existing biological systems. In a paper published in 2020, my colleagues and I showed progress in one of the key challenges with lab-grown organs: figuring out the genes necessary to produce the variety of mature cells needed to construct a functioning liver.[1]

Induced pluripotent stem cells, a subgroup of stem cells, are capable of producing cells that can build entire organs in the human body. But they can do this job only if they receive the right quantity of growth signals at the right time from their environment. If this happens, they eventually give rise to different cell types that can assemble and mature in the form of human tissues and organs. The tissues that researchers generate from pluripotent stem cells can provide a unique source for personalized medicine, from transplantation to novel drug discovery.[2]

But unfortunately, synthetic tissues from stem cells are not always suitable for transplant or drug testing because they contain unwanted cells from other tissues or they lack the maturity and a complete network of blood vessels necessary for supplying the oxygen and nutrients needed to nurture an organ. That is why it is critical to have a framework to assess whether these lab-grown cells and tissues are doing their job and to determine how to make them more like human organs.

Inspired by this challenge, I was determined to establish a synthetic biology method to read and write, or program, tissue development.[3] I am trying to do this using the genetic language of stem cells, similar to what is used by nature to form human organs.

Tissues and Organs Made by Genetic Designs

I am a researcher specializing in synthetic biology and biological engineering at the Pittsburgh Liver Research Center and McGowan Institute for Regenerative Medicine, where the goals are to use engineering approaches to analyze and build novel biological systems and solve human health problems. My lab combines synthetic biology and regenerative medicine in a new field that strives to replace, regrow, or repair diseased organs or tissues.

I chose to focus on growing new human livers because this organ is vital for controlling most levels of chemicals, like proteins or sugar, in the blood. The liver also breaks down harmful chemicals and metabolizes many drugs in our body. But liver tissue is vulnerable and can be damaged or destroyed by many diseases, such as hepatitis or fatty liver disease. There is a shortage of donor organs, which limits liver transplantation.

To make synthetic organs and tissues, scientists need to be able to control stem cells so that they can transform into different types of cells, such as liver cells and blood vessel cells. The goal is to mature these stem cells into mini-organs, or organoids,[4] containing blood vessels and the correct adult cell types that would be found in a natural organ.

One way to orchestrate the maturation of synthetic tissues is to determine the list of genes needed to induce a group of stem cells to grow and differentiate into a complete and functioning organ. To derive this list I worked with Patrick Cahan and Samira Kiani. We used computational analysis to identify genes involved in transforming a group of stem cells into a mature functioning liver.

Then our team, led by two of my students, Jeremy Velazquez and Ryan LeGraw, undertook genetic engineering to alter specific genes we had identified and used them to help build and mature human liver tissues from stem cells. The tissue is grown from a layer of genetically engineered stem cells in a petri dish. The function of genetic programs together with nutrients is to orchestrate the formation of liver organoids over the course of 15 to 17 days.

Liver in a Dish

My colleagues and I first compared the active genes in fetal liver organoids that we had grown in the lab with those in adult human livers using computational analysis; this let us generate a list of genes needed for driving fetal liver organoids to mature into adult organs.[5] We then used genetic engineering to tweak genes and the resulting proteins that the stem cells needed to mature further toward an adult liver. In the course of about 17 days we generated tiny—several millimeters in width—but more mature liver tissues with a range of cells typically found in livers in the third trimester of human pregnancies.

Like a mature human liver, these synthetic livers were able to store, synthesize, and metabolize nutrients. Though our lab-grown livers were small, we are hopeful that we can scale them up in the future. While they share many features of adult livers, they aren't perfect, and our team has work yet to do. For example, we still need to improve the capacity of the liver tissue to metabolize a variety of drugs. We also need to make it safer and more efficacious for eventual application in humans.

Our study demonstrates the ability of these lab-grown livers to mature and develop a working network of blood vessels in just two and a half weeks. The liver organoids perform several vital functions of an adult human liver such as the production of key blood proteins and the regulation of bile, a chemical important for digesting food. When we implanted the lab-grown liver tissue in mice suffering from liver disease, it increased their life span. We named our organoids "designer organoids," as they are generated through a genetic design. We believe this approach can pave the way for the manufacture of other organs with vasculature via genetic programming.

Notes

1. Velazquez, J. J., LeGraw, R., Moghadam, F., Tan, Y., Kilbourne, J., Maggiore, J. C., Hislop, J., Liu, S., Cats, D., Chuva de Sousa Lopes, S. M., Plaisier, C., Cahan, P., Kiani, S., & Ebrahimkhani, M. R. (2020). Gene regulatory network analysis and engineering directs development and vascularization of multilineage human liver organoids. *Cell Systems, 12*(1). https://doi.org/10.1016/j.cels.2020.11.002.
2. *Personalized medicine: A biological approach to patient treatment.* (2016, February 26). US Food and Drug Administration. https://www.fda.gov/drugs/news-events-human-drugs/personalized-medicine-biological-approach-patient-treatment.
3. *What is synthetic/engineering biology?* (n.d.). Engineering Biology Research Consortium. https://ebrc.org/what-is-synbio/.
4. *Organoids.* (2018, January). Nature. https://www.nature.com/articles/nmeth.4576.pdf?origin=ppub.
5. Guye, P., Ebrahimkhani, M. R., Kipniss, N., Velazquez, J. J., Schoenfeld, E., Kiani, S., Griffith, L. G., & Weiss, R. (2016). Genetically engineering self-organization of human pluripotent stem cells into a liver bud-like tissue using GATA6. *Nature Communications, 7*(1). https://doi.org/10.1038/ncomms10243.

Chapter 20

We're Creating "Humanized Pigs" in Our Ultraclean Lab to Study Human Illnesses and Treatments

CHRISTOPHER TUGGLE and ADELINE BOETTCHER

THE US FOOD AND DRUG ADMINISTRATION requires that all new medicines be tested in animals before they are used in people. Pigs make better medical research subjects than mice because they are closer to humans in size, physiology,[1] and genetic makeup.[2] In recent years, our team at Iowa State University has found a way to make pigs an even closer stand-in for humans. We have successfully transferred components of the human immune system into pigs that lack a functional immune system.[3] This breakthrough has the potential to accelerate medical research in many areas, including viruses and vaccines, as well as cancer and stem cell therapeutics.

Existing Biomedical Models

Severe combined immunodeficiency, or SCID, is a genetic condition that causes impaired development of the immune system. People can develop SCID, as was dramatized in the 1976 movie *The Boy in the Plastic Bubble*. Other animals can develop SCID, too, including mice.

Researchers in the 1980s recognized that SCID mice could be implanted with human immune cells for further study.[4] Such mice are called "humanized" mice and have been optimized over the past 30 years to study many questions relevant to human health. Mice are the most commonly used animal in biomedical research, but results from mice often do not translate well to human responses, thanks to differences in metabolism, size, and divergent cell functions compared with people.[5]

Nonhuman primates are also used for medical research and are certainly closer stand-ins for humans. But using them for this purpose raises numerous ethical considerations. With these concerns in mind, the National Institutes of Health retired most of its chimpanzees from biomedical research in 2013. That led to a demand for alternative animal models.

Swine are a viable option for medical research because of their similarities to humans. And with their widespread commercial use, pigs raise fewer ethical dilemmas than primates. Upward of 100 million hogs are slaughtered each year for food in the United States.

Humanizing Pigs

In 2012, groups at Iowa State University and Kansas State University, which included Jack Dekkers, an expert in animal breeding and genetics, and Raymond Rowland, a specialist in animal diseases, serendipitously discovered a naturally occurring genetic mutation in pigs that caused SCID. Our research team wondered if we could develop these pigs and create a new biomedical model.

Our group at ISU has worked for nearly a decade to optimize SCID pigs for applications in biomedical research. In 2018 we achieved a twofold milestone when working with animal physiologist Jason Ross and his lab. Together, we developed a more immunocompromised pig than the original SCID pig and successfully humanized it by transferring cultured human immune stem cells into the livers of developing piglets.

During early fetal development, immune cells develop in the liver, providing an opportunity to introduce human cells. We inject human immune stem cells into fetal pig livers using ultrasound imaging as a guide. As the pig fetus develops, the injected human immune stem cells begin to differentiate—or change into other kinds of cells—and spread through the pig's body. Once SCID piglets are born, we can detect human immune cells in their blood, liver, spleen, and thymus gland. This humanization is what makes them so valuable for testing new medical treatments.

We have found that human ovarian tumors survive and grow in SCID pigs, giving us an opportunity to study ovarian cancer in a new way.[6] Similarly, because human skin survives on SCID pigs, scientists may be able to develop new treatments for skin burns. Other research possibilities are numerous.

Pigs in a Bubble

Since our pigs lack essential components of their immune system, they are extremely susceptible to infection and require special housing to help reduce exposure to pathogens. SCID pigs are raised in bubble biocontainment facilities.[7] Positive-pressure rooms, which maintain a higher air pressure than the surrounding environment to keep pathogens out, are coupled with highly filtered air and water. All personnel are required to wear full personal protective equipment. We typically have anywhere from 2 to 15 SCID pigs and breeding animals at a given time. (Our breeding animals do not have SCID, but they are genetic carriers of the mutation, so their offspring may have SCID.)

As with any animal research, ethical considerations are always front and center. All our protocols are approved by Iowa State University's Institutional Animal Care and Use Committee and are in accordance with the National Institutes of Health's *Guide for the Care and Use of Laboratory Animals*. Every day, twice a day, our pigs are checked by expert caretakers who monitor their health status and provide engagement. We have veterinarians on call. If any pigs fall ill, and drug or antibiotic intervention does not improve their condition, the animals are humanely euthanized.

Our goal is to continue optimizing our humanized SCID pigs so they can be more readily available for stem cell therapy testing, as well as research in other areas, including cancer. We hope the development of the SCID pig model will pave the way for advancements in therapeutic testing, with the long-term goal of improving outcomes for human patients.

Notes

1. Meurens, F., Summerfield, A., Nauwynck, H., Saif, L., & Gerdts, V. (2012). The pig: A model for human infectious diseases. *Trends in Microbiology, 20*(1), 50–57. https://doi.org/10.1016/j.tim.2011.11.002.
2. Dawson, H. D., Loveland, J. E., Pascal, G., Gilbert, J. G. R., Uenishi, H., Mann, K. M., Sang, Y., Zhang, J., Carvalho-Silva, D., Hunt, T., Hardy, M., Hu, Z., Zhao, S.-H., Anselmo, A., Shinkai, H., Chen, C., Badaoui, B., Berman, D., Amid, C., . . . Tuggle, C. K. (2013, May 15). Structural and functional annotation of the porcine immunome. *BMC Genomics.* https://bmcgenomics.biomedcentral.com/articles/10.1186/1471-2164 -14-332.
3. Boettcher, A. N., Li, Y., Ahrens, A. P., Kiupel, M., Byrne, K. A., Loving, C. L., Cino-Ozuna, A. G., Wiarda, J. E., Adur, M., Schultz, B., Swanson, J. J., Snella, E. M., Ho, C.-S. (S.), Charley, S. E., Kiefer, Z. E., Cunnick, J. E., Putz, E. J., Dell'Anna, G., Jens, J., . . . Tuggle, C. K. (2020, February 6). Novel engraftment and T cell differentiation of human hematopoietic cells in *ART*$^{-/-}$ *IL2RG*$^{-/Y}$ SCID pigs. *Frontiers in Immunology.* https:// www.frontiersin.org/articles/10.3389/fimmu.2020.00100/full.
4. Mosier, D. E., Gulizia, R. J., Baird, S. M., & Wilson, D. B. (1988). Transfer of a functional human immune system to mice with severe combined immunodeficiency. *Nature, 335*(6187), 256–259. https://doi.org/10 .1038/335256a0.
5. Hodge, R. D., Bakken, T. E., Miller, J. A., Smith, K. A., Barkan, E. R., Graybuck, L. T., Close, J. L., Long, B., Johansen, N., Penn, O., Yao, Z., Eggermont, J., Höllt, T., Levi, B. P., Shehata, S. I., Aevermann, B., Beller, A., Bertagnolli, D., Brouner, K., . . . Lein, E. S. (2019). Conserved cell types with divergent features in human versus mouse cortex. *Nature, 573*(7772), 61–68. https://doi.org/10.1038/s41586-019-1506-7.
6. Boettcher, A. N., Kiupel, M., Adur, M., Cocco, E., Santin, A., Charley, S., Risinger, J., Tuggle, C., & Shapiro, E. (2018). Abstract LB-042: Successful tumor formation following xenotransplantation of primary human ovarian cancer cells into severe combined immunodeficient (SCID) pigs. *Tumor Biology.* https://doi.org/10.1158/1538-7445 .am2018-lb-042.
7. Powell, E. J., Charley, S., Boettcher, A. N., Varley, L., Brown, J., Schroyen, M., Adur, M. K., Dekkers, S., Isaacson, D., Sauer, M., Cunnick, J., Ellinwood, N. M., Ross, J. W., Dekkers, J. C. M., & Tuggle, C. K. (2018). Creating effective biocontainment facilities and maintenance protocols for raising specific pathogen-free, severe combined immunodeficient (SCID) pigs. *Laboratory Animals, 52*(4), 402–412. https://doi.org/10.1177/0023677217750691.

Chapter 21

When Researchers Don't Have the Proteins They Need, They Can Get AI to "Hallucinate" New Structures

IVAN ANISHCHENKO

ALL LIVING ORGANISMS USE PROTEINS, which encompass a vast number of complex molecules. They perform a wide array of functions, from allowing plants to use solar energy for oxygen production, to helping your immune system fight against pathogens, and to letting your muscles perform physical work. Many drugs are also based on proteins.[1]

For many areas of biomedical research and drug development, however, there are no natural proteins that can serve as suitable starting points to build new proteins. Researchers designing new drugs to prevent COVID-19 infection, or developing proteins that can turn genes on or off, had to create new proteins from scratch.[2] This process of de novo protein design can be difficult to get right. Protein engineers like me have been trying to figure out more efficient and accurate ways to design new proteins with the properties we need.

Luckily, a form of artificial intelligence called deep learning may provide an elegant way to create proteins that did not exist previously—hallucination.[3]

Designing Proteins from Scratch

Proteins are made up of hundreds to thousands of smaller building blocks called amino acids. These amino acids are connected to one another in long chains that fold to form a protein. The order in which these amino acids are connected to one another determines each protein's unique structure and function.

The biggest challenge protein engineers face when designing new proteins is coming up with a protein structure that will perform a desired function. To get around this problem, researchers typically create design templates based on naturally occurring proteins with a similar function. These templates have instructions for how to create the unique folds of each particular protein. However, because a template must be created for each individual fold, this strategy is time-consuming, labor-intensive, and limited by what proteins are available in nature.

Over the past few years, various research groups, including the lab I work in, have developed a number of dedicated deep neural networks: computer programs that use multiple processing layers to "learn" from input data to make predictions about a desired output. When the desired output is a new protein, millions of parameters describing different facets of a protein are put into the network.[4] What's predicted is a randomly chosen sequence of amino acids mapped onto the most probable three-dimensional structure that sequence would take. Network predictions for a random amino acid sequence are blurry, meaning the final structure of the protein is not clear-cut, but proteins that are either naturally occurring or built from scratch produce much more well-defined protein structures.

Hallucinating New Proteins

These observations hint at one way that new proteins can be generated from scratch: that is, by tweaking random inputs to the network until predictions yield a well-defined structure. The protein generation method my colleagues and I developed is conceptually similar to computer vision methods such as Google's DeepDream, which finds and enhances patterns in images.

These methods work by taking networks trained to recognize human faces or other patterns in images, like the shape of an animal or an object, and inverting them so that they learn to recognize these patterns where they don't exist. In DeepDream, for example, the network is given arbitrary input images that are adjusted until the network can recognize a face or some other shape in the image. While the final image doesn't look much like a face to a person looking at it, it would to the neural network.

The products of this technique are often referred to as hallucinations, and this is what we call our designed proteins, too.

Our method starts by passing a random amino acid sequence through a deep neural network. The resulting predictions are initially blurry, with unclear structures, as expected for random sequences. Next, we introduce a mutation that changes one amino acid in the chain into a different one and pass this new sequence through the network again. If this change gives the protein a more defined structure, then we keep the amino acid and introduce another mutation into the sequence. With each repetition of this process, the proteins get closer and closer to the real shape they would take if they were produced in nature. Thousands of repetitions are required to create a brand-new protein.

Using this process, we generated 2,000 new protein sequences predicted to fold into well-defined structures. Of these, we selected over 100 that were the most distinctive in shape to physically re-create in the lab. Finally, we chose three of the top candidates for detailed analysis and confirmed that they were close matches to the shapes predicted by our hallucinated models.

Why Hallucinate New Proteins?

Our hallucination approach greatly simplifies the protein design pipeline. By eliminating the need for templates, researchers can focus directly on creating a protein with the desired functions and let the network take care of figuring out the structure for them.

Our work opens up multiple avenues for researchers to explore. Our lab is investigating how best to use this hallucination approach to generate even more specificity in the function of designed proteins.[5] Our approach can also be readily extended to design new proteins using other deep neural networks. The potential applications of de novo proteins are vast: with deep neural networks, researchers will be able to create more proteins that break down plastics to reduce environmental pollution, that identify and respond to unhealthy cells, and that improve vaccines for existing and new pathogens[6]—just to name a few.

Notes

1. Dimitrov, D. S. (2012). Therapeutic proteins. In V. Voynov & Caravella (Eds.). *Therapeutic proteins: Methods and protocols.* (2nd ed., pp. 1–26). Humana Press. https://doi.org/10.1007/978-1-61779-921-1_1.

2. Cao, L., Goreshnik, I., Coventry, B., Case, J. B., Miller, L., Kozodoy, L., Chen, R. E., Carter, L., Walls, A. C., Park, Y.-J., Strauch, E.-M., Stewart, L., Diamond, M. S., Veesler, D., & Baker, D. (2020). De novo design of picomolar SARS-CoV-2 miniprotein inhibitors. *Science, 370*(6515), 426–431. https://doi.org/10.1126/science.abd9909.

3. Anishchenko, I., Pellock, S. J., Chidyausiku, T. M., Ramelot, T. A., Ovchinnikov, S., Hao, J., Bafna, K., Norn, C., Kang, A., Bera, A. K., DiMaio, F., Carter, L., Chow, C. M., Montelione, G. T., & Baker, D. (2021). De novo protein design by deep network hallucination. *Nature, 600*(7889), 547–552. https://doi.org/10.1038/s41586-021-04184-w.

4. Yang, J., Anishchenko, I., Park, H., Peng, Z., Ovchinnikov, S., & Baker, D. (2020). Improved protein structure prediction using predicted interresidue orientations. *Proceedings of the National Academy of Sciences, 117*(3), 1496–1503. https://doi.org/10.1073/pnas.1914677117.

5. Wang, J., Lisanza, S., Juergens, D., Tischer, D., Watson, J. L., Castro, K. M., Ragotte, R., Saragovi, A., Milles, L. F., Baek, M., & Anishchenko, I. (2022). Scaffolding protein functional sites using deep learning. *Science, 377*(6604), 387–394. https://doi.org/10.1126/science.abn2100.

6. Sesterhenn, F., Yang, C., Bonet, J., Cramer, J. T., Wen, X., Wang, Y., Chiang, C.-I., Abriata, L. A., Kucharska, I., Castoro, G., Vollers, S. S., Galloux, M., Dheilly, E., Rosset, S., Corthésy, P., Georgeon, S., Villard, M., Richard, C.-A., Descamps, D., . . . Correia, B. E. (2020). De novo protein design enables the precise induction of RSV-neutralizing antibodies. *Science, 368*(6492). https://doi.org/10.1126/science.aay5051.

Chapter 22

Living Drugs: Engineering Bacteria to Treat Genetic Diseases

PEDRO BELDA-FERRE

A PILL CONTAINING MILLIONS OF BACTERIA ready to colonize your gut might sound a nightmare to many. But it may become an effective new tool for fighting disease. In many inherited genetic diseases, a mutated gene means that an individual cannot make a vital substance necessary for their body to grow, develop, or function. Sometimes this can be fixed with a synthetic substitute, a pill, which they can take daily to replace what their body should have made naturally.

People with a rare genetic disease called phenylketonuria (PKU) lack an enzyme that is essential for breaking down protein. Without it, toxic chemicals build up in the blood and can cause permanent brain damage. Fortunately, the fix is easy. Physicians treat the disease by putting their patients on a super-low-protein diet for the rest of their life. Indeed, because the fix was so simple, PKU was the first disorder for which newborn babies were routinely screened, beginning in 1961, by analyzing a drop of blood collected from a prick on the baby's heel. But imagine how challenging it can be to monitor everything you eat for your entire life.

To cure PKU, researchers are currently exploring new treatment strategies. One involves using gene-editing tools to correct genetic mutations. However, the current technology is still risky, as there is a chance of disrupting other genes and causing collateral damage to patients.[1] What if one could instead replace the broken gene without affecting the patient's genome? That's exactly what researchers at Syn-logic, a biotech company based in Cambridge, Massachusetts, have done. They decided that rather than meddle directly with the human genome, they would introduce the therapeutic genes directly into the naturally occurring bacteria that reside in the human gut. These genetically modified bacteria would then produce the enzymes that PKU patients lacked and break down the proteins into nontoxic products.

I am an assistant project scientist at the University of California, San Diego, who studies the community of microbes that live in our bodies to learn how they impact our health. We are starting to understand now the part these microbes play in maintaining our health. The next step is figuring out how we can alter them to improve our health. And Synlogic's research is bringing that dream a step closer to reality.

Engineering Bacteria Living in Our Gut

You may be surprised to learn that our intestines are inhabited by trillions of bacteria that help us digest food,[2] produce vitamins for us,[3] and educate our immune system.[4] This community of microbes is called our microbiome. Together, these microbes harbor millions of different genes in their genomes, outnumbering our human genes 150 to 1, and we can use them to our own benefit.

Escherichia coli Nissle 1917 is one of those microbes living inside most of us and has been widely used as a probiotic for over a century and has proven its safety.[5] It is what Synlogic chose to engineer as a new therapeutic "super bacterium" called SYNB1618 for the benefit of PKU patients.

Synlogic's researchers introduced three genes that enable SYNB1618 to transform one of the building blocks of protein, an amino acid called phenylalanine, into the safe compound phenylpyruvate. As long as the levels of phenylalanine are kept low, PKU patients don't show any symptoms and live normal lives.

Are Genetically Modified Bacteria Safe?

Opponents of genetically modified organisms might object to inserting designer microbes into our guts. But just as with genetically modified foods, strict FDA regulations govern the safety of these experimental microbes.

In the case of SYNB1618, researchers deleted a gene responsible for producing an essential ingredient in building the bacterium. If researchers don't provide the missing ingredient for the engineered bacterium, it can't replicate and will die. This mechanism is a way for researchers to control

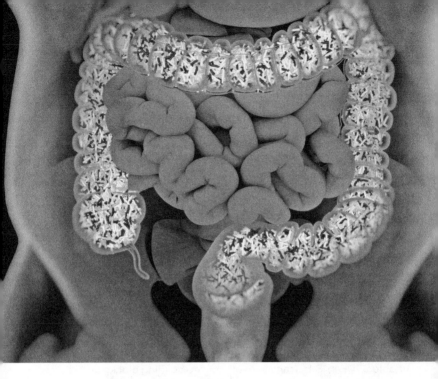

SYNB1618 in a patient's body. When they tested the microbe in mice, they discovered that after 48 hours without the crucial ingredient, SYNB1618 had vanished from their guts. The researchers at Synlogic took other precautions, as well, in engineering SYNB1618. The engineered bacterium contains exactly the same genes as the original *E. coli* Nissle 1917 that is native to the gut, thus ensuring the safety of SYNB1618.

Does It Really Work?

Once the researchers had proven that the bacteria could convert phenylalanine in the lab, they decided to administer the bacteria to mice with PKU. The results showed that

The large intestine is filled with a community of bacteria and other microorganisms that are essential for good health. One of these naturally occurring gut bacteria was re-engineered by Synlogic and altered to help treat the genetic disease phenylketonuria. *Anatomy Insider / Shutterstock.com*

SYNB1618 degraded phenylalanine circulating in the animals' intestines, which in turn lowered the levels of phenylalanine in the blood of the treated mice. Then, in preparing for tests in humans, researchers tested SYNB1618 in monkeys. Healthy monkeys without PKU were fed phenylalanine and given a dose of the microbes afterward. The SYNB1618 bacteria successfully reduced the phenylalanine blood levels—just as they had in mice.

Synlogic is currently testing SYNB1618 in humans in clinical trials. This is a step toward a new therapeutic approach that offers great potential to treat human diseases like diabetes and cancer and to monitor inflammation levels in inflammatory bowel diseases.[6]

As we discover and understand the role of all the microbes that inhabit our bodies, I expect that we will identify microbes that may be the perfect vehicles for carrying various gene therapies that treat even more diseases, including those involving metabolism and the central nervous system.

Notes

1. Peng, R., Lin, G., & Li, J. (2015). Potential pitfalls of CRISPR/Cas9-mediated genome editing. *FEBS Journal*, *283*(7), 1218–1231. https://doi.org/10.1111/febs.13586.
2. Sonnenburg, J. L., Xu, J., Leip, D. D., Chen, C.-H., Westover, B. P., Weatherford, J., Buhler, J. D., & Gordon, J. I. (2005). Glycan foraging in vivo by an intestine-adapted bacterial symbiont. *Science*, *307*(5717), 1955–1959. https://doi.org/10.1126/science.1109051.

3. Karl, J. P., Meydani, M., Barnett, J. B., Vanegas, S. M., Barger, K., Fu, X., Goldin, B., Kane, A., Rasmussen, H., Vangay, P., Knights, D., Jonnal-agadda, S. S., Saltzman, E., Roberts, S. B., Meydani, S. N., & Booth, S. L. (2017). Fecal concentrations of bacterially derived vitamin K forms are associated with gut microbiota composition but not plasma or fecal cytokine concentrations in healthy adults. *American Journal of Clinical Nutrition*, *106*(4), 1052–1061. https://doi.org/10.3945/ajcn.117.155424.

4. Olszak, T., An, D., Zeissig, S., Vera, M. P., Richter, J., Franke, A., Glickman, J. N., Siebert, R., Baron, R. M., Kasper, D. L., & Blumberg, R. S. (2012). Microbial exposure during early life has persistent effects on natural killer T cell function. *Science*, *336*(6080), 489–493. https://doi.org/10.1126/science.1219328.

5. Sonnenborn, U. (2016). *Escherichia coli* strain Nissle 1917—from bench to bedside and back: History of a special *Escherichia coli* strain with probiotic properties. *FEMS Microbiology Letters*, *363*(19). https://doi.org/10.1093/femsle/fnw212.

6. Riglar, D. T., Giessen, T. W., Baym, M., Kerns, S. J., Niederhuber, M. J., Bronson, R. T., Kotula, J. W., Gerber, G. K., Way, J. C., & Silver, P. A. (2017). Engineered bacteria can function in the mammalian gut long-term as live diagnostics of inflammation. *Nature Biotechnology*, *35*(7), 653–658. https://doi.org/10.1038/nbt.3879.

Chapter 23

How Gene-Editing a Person's Brain Cells Could Be Used to Curb the Opioid Epidemic

CRAIG W. STEVENS

THERE IS AN ONGOING EPIDEMIC killing tens of thousands of people each year: opioid drug overdose.[1] Opioid analgesic drugs, like morphine and oxycodone, are the classic double-edged sword: they are the best drugs to stop severe pain, but they also belong to a class of drugs with the potential to accidentally kill the person taking them.

I am a pharmacologist interested in the way that opioid drugs such as morphine and fentanyl can blunt pain. I became fascinated with biology at the time when endorphins—natural opioids made by our bodies—were discovered. I have been intrigued by the way opioid drugs act on their targets in the brain, the opioid receptors, for the last 30 years. In a publication, I proposed a way of combining state-of-the-art molecular techniques, such as CRISPR gene editing and brain microinjection, to blunt one edge of the opioid sword and so prevent fatal overdoses.[2]

Opioid Receptors Stop Breathing

Opioids kill by stopping a person from breathing (respiratory depression). They do so by acting on a specific set of respiratory nerves, or neurons, in the lower part of the brain that contain opioid receptors. Opioid receptors are proteins that bind morphine, heroin, and other opioid drugs. The binding of an opioid to its receptor triggers a reaction in neurons that reduces their activity. Opioid receptors on pain neurons mediate the painkilling, or analgesic, effects of opioids. When opioids bind to opioid receptors on respiratory neurons, they slow breathing or, in the case of an opioid overdose, stop it entirely.

Respiratory neurons are located in the brainstem, the tail-end of the brain that continues into the spine as the spinal cord. Animal studies show that opioid receptors on respiratory neurons are responsible for opioid-induced respiratory depression,[3] the result of an opioid overdose. Genetically altered mice born without opioid receptors, however, do not die from large doses of morphine,[4] in contrast with normal mice whose brain cells have these receptors.

Unlike laboratory mice, humans cannot be altered when embryos to remove all opioid receptors from the brain and elsewhere. Nor would it be a good idea. Humans need opioid receptors to serve as the targets for our natural opioid substances, the endorphins, which are released into the brain during times of high stress and pain. A total opioid receptor knockout would leave a person unresponsive to the beneficial pain-killing effects of opioids.

In my journal article, I argue that what is needed is a selective receptor removal of the opioid receptors on respiratory neurons. Having reviewed the available technology, I believe this can be done by combining CRISPR gene editing and a new neurosurgical microinjection technique.

CRISPR to the Rescue: Destroying Opioid Receptors

CRISPR, which is an acronym for clustered regularly interspaced short palindromic repeats, is a gene-editing method that was discovered in the genome of bacteria. Bacteria get infected by viruses too, and CRISPR is a strategy that bacteria evolved to cut up the viral genes and kill invading pathogens.

The CRISPR method allows researchers to target specific genes expressed in cell lines, tissues, or whole organisms and then cut them up and remove them—that is, knock them out—or otherwise alter them. There is a commercially available CRISPR kit that knocks out human opioid receptors produced in cells grown in cell cultures in the lab. While this CRISPR kit is formulated for in vitro use, similar conditional opioid receptor knockout techniques have been demonstrated in live mice.[5]

To knock out opioid receptors in human respiratory neurons, a sterile solution containing CRISPR gene-editing molecules would be prepared in the laboratory. Besides the gene-editing components, the solution would contain chemical reagents that allow the gene-editing machinery to enter the respiratory neurons and make their way into the nucleus and then into the neuron's genome.

But how does one get the CRISPR opioid receptor knock-out solution into a person's respiratory neurons specifically?

Enter the intracranial microinjection instrument (IMI) developed by Miles Cunningham and his colleagues at Harvard University. The IMI allows for computer-controlled delivery of small volumes of solution at specific places in the brain by using an extremely thin tube—about twice the diameter of a human hair—that can be inserted into the brain at the base of the skull and threaded through brain tissue without damage. Because the brain itself feels no pain, the procedure could be done in a conscious patient using only local anesthetic to numb the skin.

A computer can direct the robotic placement of the tube when it is fed images of the brain taken before the procedure using MRI, a type of internal imaging. But even better, the IMI also has a recording wire embedded in the tube that measures neuronal activity to identify the right group of nerve cells. Respiratory neurons drive the breathing muscles by firing action potentials, which are measured by the recording wire in the tube. When the activity of the respiratory neurons matches the breathing movements of the patient, the proper location of the tube is confirmed, and the CRISPR solution is injected.

The Call for Drastic Action

Opioid receptors on neurons in the brain have a half-life of about 45 minutes.[6] Over a period of several hours, the opioid receptors on respiratory neurons would degrade, and the CRISPR gene-editing machinery embedded in the genome would prevent new opioid receptors from appearing. If this works, the patient would be protected from opioid overdose within 24 hours. Because the respiratory neurons do not replenish, the CRISPR opioid receptor knockout should last for life.

With no opioid receptors on respiratory neurons, the opioid user cannot die from opioid overdose. With proper backing from the National Institute on Drug Abuse and leading research and health care institutions, I believe CRISPR treatment of this kind could enter clinical trials between 5 and 10 years down the road. The total cost of opioid-involved overdose deaths is about 430 billion US dollars per year. CRISPR treatment for only 10 percent of high-risk opioid users in one year would save thousands of lives and 43 billion dollars.

Intracranial microinjection of CRISPR solutions might seem drastic. But drastic actions are needed to save human lives from opioid overdoses. A large segment of opioid overdose victims is patients with chronic pain.[7] It may be possible that chronic pain patients in a terminal phase of their lives and in hospice care would volunteer for phase I clinical trials of the CRISPR opioid receptor knockout treatment that I have described.

Making the opioid user impervious to death by opioid overdose is a permanent solution to a horrendous problem that has resisted prevention efforts by psychotherapeutic and pharmacological means. Careful and well-funded work to prove the CRISPR method, first with preclinical animal models and then in clinical trials, is a moonshot for the present generation of biomedical scientists.

Notes

1. Jalal, H., Buchanich, J. M., Roberts, M. S., Balmert, L. C., Zhang, K., & Burke, D. S. (2018). Changing dynamics of the drug overdose epidemic in the United States from 1979 through 2016. *Science, 361*(6408). https://doi.org/10.1126/science.aau1184.

2. Stevens, C. W. (2020). Receptor-centric solutions for the opioid epidemic: Making the opioid user impervious to overdose death. *Journal of Neuroscience Research, 100*(1), 322–328. https://doi.org/10.1002/jnr.24636.

3. Levitt, E. S., Abdala, A. P., Paton, J. F., Bissonnette, J. M., & Williams, J. T. (2015). μ opioid receptor activation hyperpolarizes respiratory-controlling Kölliker-fuse neurons and suppresses post-inspiratory drive. *Journal of Physiology, 593*(19), 4453–4469. https://doi.org/10.1113/jp270822.

4. Loh, H. H., Liu, H.-C., Cavalli, A., Yang, W., Chen, Y.-F., & Wei, L.-N. (1998). μ opioid receptor knockout in mice: Effects on ligand-induced analgesia and morphine lethality. *Molecular Brain Research, 54*(2), 321–326. https://doi.org/10.1016/s0169-328x(97)00353-7.

5. Varga, A. G., Reid, B. T., Kieffer, B. L., & Levitt, E. S. (2019). Differential impact of two critical respiratory centres in opioid-induced respiratory depression in awake mice. *Journal of Physiology, 598*(1), 189–205. https://doi.org/10.1113/jp278612.

6. Medrano, M. C., Santamarta, M. T., Pablos, P., Aira, Z., Buesa, I., Azkue, J. J., Mendiguren, A., & Pineda, J. (2017). Characterization of functional μ opioid receptor turnover in rat locus coeruleus: An electrophysiological and immunocytochemical study. *British Journal of Pharmacology, 174*(16), 2758–2772. https://doi.org/10.1111/bph.13901.

7. Bonnie, R. J., Kesselheim, A. S., & Clark, D. J. (2017). Both urgency and balance needed in addressing opioid epidemic. *JAMA, 318*(5), 423. https://doi.org/10.1001/jama.2017.10046.

Chapter 24

CRISPR Can Help Combat the Troubling Immune Response against Gene Therapy

SAMIRA KIANI

ONE OF THE MAJOR CHALLENGES facing gene therapy—a way to treat disease by replacing a patient's defective genes with healthy ones—is that it is difficult to deliver a therapeutic gene safely to patients without the immune system destroying the gene, and the vehicle carrying it, which can trigger life-threatening widespread inflammation.[1]

Three decades ago researchers thought that gene therapy would be the ultimate treatment for genetically inherited diseases like hemophilia, sickle cell anemia, and genetic diseases of metabolism. But the technology couldn't dodge the immune response. Since then, researchers have been looking for ways to perfect the technology and control immune responses to the gene or to the vehicle. However, many of the strategies tested so far have not been completely successful in overcoming this hurdle.[2]

Drugs that suppress the whole immune system, such as steroids, have been used to dampen the immune response upon administering gene therapy. But it's difficult to control when and where steroids work in the body, and they create unwanted side effects. My colleague Mo Ebrahimkhani and I wanted to attempt gene therapy with immune-suppressing tools that were easier to control.

I am a medical doctor and synthetic biologist who is interested in gene therapy because six years ago my father was diagnosed with pancreatic cancer. Pancreatic cancer is one of the deadliest forms of cancer, and the currently available therapeutics usually fail to save patients. Consequently, novel treatments such as gene therapy might be the only hope. Many gene therapies fail, however, because patients either have preexisting immunity to the vehicle used to introduce the gene or develop immunity in the course of therapy. This problem has plagued the field for decades and inhibited the widespread application of the technology.

Gene Therapy: Past and Present

Traditionally, scientists have use viruses—from which dangerous disease-causing genes have been removed—as vehicles for transporting new genes to specific organs. These genes then produce a product that can compensate for the faulty genes that were inherited genetically. This is how gene therapy works.

Although there have been examples of gene therapy helping to control genetic diseases,[3] the method is not perfected. Sometimes, a faulty gene is so big that you simply can't fit the healthy replacement in the viruses commonly used in gene therapy. Another problem is that when the immune system detects a virus, it assumes the virus is a disease-causing pathogen and launches an attack to fight it off by producing antibodies—just like what happens when people catch any other harmful virus, such as SARS-CoV-2 or a rhinovirus causing the common cold.

Recently, though, with the rise of a gene-editing technology called CRISPR, scientists can do gene therapy differently.[4]

CRISPR can be used in many ways. In its primary role, it acts like a genetic surgeon with a sharp scalpel, enabling scientists to find a genetic defect and correct it within the native genome in desired cells of the organism. CRISPR can repair more than one gene at a time. Scientists can also use CRISPR to turn off a gene for a short period of time and then turn it back on, or vice versa, without permanently changing the sequence of DNA making up a genome. This means that researchers like me can leverage CRISPR technology to revolutionize gene therapies in the coming decades.

But to use CRISPR for either of these functions, it must be packaged in a virus to get it into the human body. And the problem of preventing the immune response to gene therapy viruses still needs to be solved for CRISPR-based gene therapies to advance. So I teamed up with Ebrahimkhani to test whether we could use CRISPR to shut down a gene

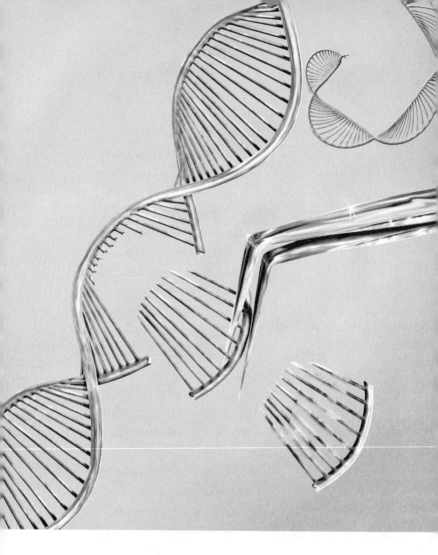

responsible for the immune response that destroys the gene therapy viruses. Then we investigated whether lowering the activity of the gene, and dulling the immune response, would allow the gene therapy viruses to be more effective.

CRISPR can precisely remove even single units of DNA.
KEITH CHAMBERS / SCIENCE PHOTO LIBRARY / Getty Images

Preventing the Immune Response That Destroys Gene Therapy Viruses

A gene called *Myd88* is a key gene of the immune system that controls the response to bacteria and viruses, including the common gene therapy viruses. We decided to temporarily turn off this gene in the whole body of lab animals.

We injected animals with a collection of CRISPR molecules that targeted the *Myd88* gene and looked to see whether this reduced the quantity of antibodies produced to fight our gene therapy viruses. We were excited to see that the animals having received our CRISPR treatment produced less antibody against the virus.

This prompted us to ask what would happen if we gave the animals a second dose of the gene therapy virus. Usually the immune response to a gene therapy virus prevents the therapy from being administered multiple times. That's because, after the first dose, the immune system has encountered the virus, so, upon the second dose, antibodies swiftly attack and destroy the virus before it can deliver its cargo. We saw, however, that animals receiving more than one dose did not show an increase in antibodies against the virus. And, in some cases, the effect of gene therapy was improved in test animals compared with control animals in which we had not paused the *Myd88* gene.

We also did a number of other experiments demonstrating that tweaking the *Myd88* gene can be useful in fighting off other sources of inflammation. That application could be useful in treating diseases like sepsis or even COVID-19.

We are now working to improve this strategy of controlling the activity of the *Myd88* gene. Our results, published in *Nature Cell Biology*, provide a path forward for programming the human immune system and other inflammatory responses during gene therapies using CRISPR technology.[5]

Notes

1. Shirley, J. L., de Jong, Y. P., Terhorst, C., & Herzog, R. W. (2020). Immune responses to viral gene therapy vectors. *Molecular Therapy*, *28*(3), 709–722. https://doi.org/10.1016/j.ymthe.2020.01.001.
2. Rinde, M. (2019, July 16). *The death of Jesse Gelsinger, 20 years later*. Science History Institute. https://www.sciencehistory.org/distillations /the-death-of-jesse-gelsinger-20-years-later.
3. Mulholland, E. J. (2020, April 17). *Start your engines: The hemophilia drug race is on*. American Society of Gene and Cell Therapy. https:// asgct.org/research/news/april-2020/world-hemophilia-day.
4. Saey, T. H. (2020, October 9). *Explainer: How CRISPR works*. Science News for Students. https://www.sciencenewsforstudents.org/article /explainer-how-crispr-works.
5. Moghadam, F., LeGraw, R., Velazquez, J. J., Yeo, N. C., Xu, C., Park, J., Chavez, A., Ebrahimkhani, M. R., & Kiani, S. (2020). Synthetic immuno-modulation with a CRISPR super-repressor in vivo. *Nature Cell Biology*, *22*(9), 1143–1154. https://doi.org/10.1038/s41556-020 -0563-3.

Chapter 25

3D-Printed Organs Could Save Lives by Addressing the Transplant Shortage

SAMAN NAGHIEH

DUE TO A GLOBAL ORGAN SHORTAGE and limited organ donors, thousands of patients are left wanting organs and tissues in cases of severe injuries, illness, or genetic conditions. Many of these patients die before a transplant becomes available.[1] Tissue engineering is an emerging field that aims to produce artificial tissue and organ substitutes as permanent solutions to replace or repair damage in the body.

As biomedical engineering researchers, we are developing three-dimensional temporary organ structures—called scaffolds—that may help regenerate damaged tissues and potentially lead to creating artificial organs. These tissues can also be used in various tissue-engineering applications, including nerve repair with structures constructed from biomaterials.

Printing Tissue

Approximately 22.6 million patients require neurosurgical interventions annually around the world to treat damage to the peripheral nervous system.[2] This damage is primarily caused by traumatic events such as motor vehicle accidents, violence, workplace injuries, or difficult births. It is anticipated that the cost of global nerve repair and regeneration will reach more than $400 million by 2025. Current surgical techniques allow surgeons to realign damaged nerve ends and encourage nerve growth. However, the incidence of recovery in the injured nervous system is not guaranteed, and the return of function is almost never complete.

Animal studies on rats have shown that if an injury destroys more than two centimeters of a nerve, the gap cannot be bridged properly and may result in the loss of muscle function or feeling.[3] In this condition, it is important to use a scaffold to bridge the two sides of the damaged nerve, specifically in a case of large nerve injuries.

Three-dimensional bioprinting prints 3D structures layer by layer, similar to 3D printers. Using this technique, our research team created a porous structure made of the patient's neural cells and a biomaterial to bridge an injured nerve. We used alginate—derived from algae—because the human body does not reject it. This technique has not yet been tested in people, but, once refined, it has the potential to help patients who are waiting for tissues or organs.

Material Challenges

Alginate is a challenging material to work with because it collapses easily during 3D printing. Our research focuses on the development of new techniques to improve its printability. For nerve repair, alginate has favorable properties for living cells' growth and function, but its poor 3D printability considerably limits its fabrication. It means that alginate flows easily during the printing process, which can result in a collapsed structure. We developed a fabrication method whereby cells are contained within a porous alginate structure created with a 3D printer.[4]

Previous research used molding techniques to create a bulk alginate without a porous structure to improve nerve regeneration; however, cells do not like such a solid environment. Alternatively, 3D-printing a porous alginate structure is challenging and often impossible. Our research addresses this issue by printing a porous structure made of alginate layer-by-layer rather than a molded bulk alginate; such a structure has interconnected pores and provides a cell-friendly environment. Cells can easily communicate with each other and start the regeneration while the 3D-printed alginate provides a temporary support for them.

Researchers are working toward the implementation of 3D-printed structures for patients who suffer from nerve injuries as well as other injuries.[5] After the fabricated alginate structure is implanted in a patient, the big question is whether it will have enough mechanical stability to tolerate the forces applied by tissues in the body. We developed a novel numerical model to predict the mechanical behavior of alginate structures.[6] Our studies are contributing to an understanding of cell response, which is the main factor to take into account when evaluating the success of the alginate structures.

Notes

1. Gridelli, B., & Remuzzi, G. (2000). Strategies for making more organs available for transplantation. *New England Journal of Medicine*, *343*(6), 404–410. https://doi.org/10.1056/nejm200008103430606.
2. Sarker, M. D., Naghieh, S., McInnes, A. D., Schreyer, D. J., & Chen, X. (2018). Regeneration of peripheral nerves by nerve guidance conduits: Influence of design, biopolymers, cells, growth factors, and physical stimuli. *Progress in Neurobiology*, *171*, 125–150. https://doi.org/10.1016/j.pneurobio.2018.07.002; Dewan, M. C., Rattani, A., Fieggen, G., Arraez, M. A., Servadei, F., Boop, F. A., Johnson, W. D., Warf, B. C., & Park, K. B. (2019). Global neurosurgery: The current capacity and deficit in the provision of essential neurosurgical care. Executive summary of the global neurosurgery initiative at the Program in Global Surgery and Social Change. *Journal of Neurosurgery*, *130*(4), 1055–1064. https://doi.org/10.3171/2017.11.jns171500.
3. Sinis, N., Schaller, H.-E., Schulte-Eversum, C., Schlosshauer, B., Doser, M., Dietz, K., Rösner, H., Müller, H.-W., & Haerle, M. (2005). Nerve regeneration across a 2-cm gap in the rat median nerve using a resorbable nerve conduit filled with Schwann cells. *Journal of Neurosurgery*, *103*(6), 1067–1076. https://doi.org/10.3171/jns.2005.103.6.1067.
4. Naghieh, S., Sarker, M. D., Abelseth, E., & Chen, X. (2019). Indirect 3D bioprinting and characterization of alginate scaffolds for potential nerve tissue engineering applications. *Journal of the Mechanical Behavior of Biomedical Materials*, *93*, 183–193. https://doi.org/10.1016/j.jmbbm.2019.02.014.
5. Sarker, M. D., Naghieh, S., McInnes, A. D., Schreyer, D. J., & Chen, X. (2018). Regeneration of peripheral nerves by nerve guidance conduits: Influence of design, biopolymers, cells, growth factors, and physical stimuli. *Progress in Neurobiology*, *171*, 125–150. https://doi.org/10.1016/j.pneurobio.2018.07.002.
6. Naghieh, S., Karamooz-Ravari, M. R., Sarker, M. D., Karki, E., & Chen, X. (2018). Influence of crosslinking on the mechanical behavior of 3D printed alginate scaffolds: Experimental and numerical approaches. *Journal of the Mechanical Behavior of Biomedical Materials*, *80*, 111–118. https://doi.org/10.1016/j.jmbbm.2018.01.034.

Chapter 26

From Marmots to Mole-Rats to Marmosets—Studying Many Genes in Many Animals Is Key to Understanding How Humans Can Live Longer

AMANDA KOWALCZYK

MUCH OF LONGEVITY AND AGING RESEARCH focuses on studying extremely long-lived species, including bats, naked mole-rats, and bowhead whales, to find genetic changes that contribute to long life.

However, such work has yielded highly species-specific genetic changes that are not generalizable to other species, including humans. As a PhD student, I studied growing evidence, including recent work from the labs of my advisors

(Maria Chikina and Nathan Clark), that supports the hypothesis that life span is a complex and highly context-dependent trait, which calls for a shift in how biologists think about aging.

Old Age: The Human Problem

Aging is the process by which the likelihood of death increases the longer an organism is alive. In mammals, aging is hallmarked by several molecular changes, including the breakdown of DNA, a shortage of stem cells, and malfunctioning proteins.[1]

The numerous theories that explain why aging happens fall into two categories. On the one hand, "wear and tear" theories postulate that essential processes simply wear out over time. On the other hand, "programmed death" theories assert that specific genes or processes are designed to drive aging. Traditional aging theories are human-centric, and when we examine aging from a cross-species perspective, it becomes clear that human aging is unique. In fact, among animals there is no typical way to age.[2]

Humans show low mortality rates until a sharp spike in mortality at very old age, around 80 years. Most mammals have relatively less increase in mortality with age and more consistent mortality through their life spans. Some mammals, such as the tundra vole and the yellow-bellied marmot, show virtually no increase in mortality with age. In other words, older individuals are equally as likely to die as younger individuals, possibly because aging does not impact survival.

Current aging theories fail to explain the complexity of aging across all mammals, let alone the tree of life. Such diversity not only highlights the complexity of aging and longevity but also makes it difficult to apply knowledge gained about one species to increase life span in another.

An Overabundance of "Longevity Genes"

Studies of exceptionally long-lived species have produced a plethora of so-called longevity genes. One such gene, a gene called *insulin-like growth factor 1 receptor*, or *IGF1R*, promotes cell growth. *IGF1R* was originally associated with long life in bats and also increases life span in worms and mice. *IGF1R* may have the opposite effect in humans, however, because too much *IGF1R* may increase age-related illnesses like diabetes and cancer.[3]

Another potential longevity gene abbreviated *ERCC1* produces a protein that helps repair DNA. The bowhead whale, the longest-lived mammal at 211 years (the oldest specimen in one sample), has a mutation in the *ERCC1* gene that may contribute to the species' exceptionally long life span, but the mutation is not shared by other long-lived species.[4] Elephants have 19 copies of the *TP53* gene, essential to cancer prevention,[5] but adding even one extra *TP53* gene to mice accelerates aging because stem cells are slower to regenerate.[6]

Longevity genes can be inconsistent even within a single species. Studies that hunt for genetic changes common in long-lived humans, and absent from humans who lived shorter lives, have not delivered a master longevity gene. The genes detected are largely inconsistent across studies, as the studies relied heavily on the subpopulation of humans sampled and the definition chosen for "exceptionally long-lived."

So How Do We Find Longevity Genes?

My work supports the argument that aging researchers should not be looking for individual longevity genes.[7] Rather, biologists should be seeking many genes with similar functions working together to control longevity. Furthermore, an effective search should not just focus on a single species but on many, so as to avoid species-specific factors.

As part of a research study, I used genomes from 61 mammals to detect genes that evolved in tandem with the evolution of extreme life span, thereby uncovering longevity-related changes universal across all mammals. At the gene level, I found few longevity genes, which makes sense in light of previous work. There is probably no single gene in all mammals that regulates life span.

When I looked at the big picture, however, and considered groups of genes working together to perform a similar function, I found a strong association between longevity and pathways related to controlling cancer, examples of which are genes involved in regulating the cell cycle and programmed cell death, and pathways for immune function and DNA repair. My work highlights the importance of a new perspective on aging and longevity.

Species-specific and human genome–wide association studies have limitations that may be enriched by a broader analysis, both in terms of the genomic elements studied and the species considered. Rather than searching for a single gene in a single species that drives increased life span, broadening the search to many genes across many species can bring new insights.

Comparative studies like mine that interrogate genetic similarities and differences across long-lived species have repeatedly demonstrated the power to detect longevity-related genetic changes spread over many genes and shared by many species.[8]

While there may be no proverbial fountain of youth hidden in our genome—the one gene that multiplies our healthy years of life—scientists like me are continually improving our strategies to study longevity so that we can someday all lead longer, healthier lives.

Notes

1. López-Otín, C., Blasco, M. A., Partridge, L., Serrano, M., & Kroemer, G. (2013). The hallmarks of aging. *Cell*, *153*(6), 1194–1217. https://doi.org/10.1016/j.cell.2013.05.039
2. Jones, O. R., Scheuerlein, A., Salguero-Gómez, R., Camarda, C. G., Schaible, R., Casper, B. B., Dahlgren, J. P., Ehrlén, J., García, M. B., Menges, E. S., Quintana-Ascencio, P. F., Caswell, H., Baudisch, A., & Vaupel, J. W. (2014). Diversity of ageing across the tree of life. *Nature*, *505*, 169–173 (2014). https://doi.org/10.1038/nature12789.
3. Junnila, R. K., List, E. O., Berryman, D. E., Murrey, J. W., & Kopchick, J. J. (2013). The GH/IGF-1 axis in ageing and longevity. *Nature Reviews Endocrinology*, *9*(6), 366–376. https://doi.org/10.1038/nrendo.2013.67.
4. Keane, M., Semeiks, J., Webb, A. E., Li, Y. I., Quesada, V., Craig, T., Madsen, L. B., van Dam, S., Brawand, D., Marques, P. I., Michalak, P., . . . de Magalhães. (2015). Insights into the evolution of longevity from the bowhead whale genome. *Cell Reports*, *10*(1), 112–122. https://doi.org/10.1016/j.celrep.2014.12.008.
5. Sulak, M., Fong, L., Mika, K., Chigurupati, S., Yon, L., Mongan, N. P., Emes, R. D., & Lynch, V. J. (2016). *TP53* copy number expansion is associated with the evolution of increased body size and an enhanced DNA damage response in elephants. eLife, *5*, e11994. https://doi.org/10.7554/eLife.11994.
6. Dumble, M., Moore, L., Chambers, S. M., Geiger, H., Van Zant, G., Goodell, M. A., & Donehower, L. A. (2006). The impact of altered p53 dosage on hematopoietic stem cell dynamics during aging. *Blood*, *109*(4), 1736–1742. https://doi.org/10.1182/blood-2006-03-010413.
7. Kowalczyk, A., Partha, R., Clark, N. L., & Chikina, M. (2020). Pan-mammalian analysis of molecular constraints underlying extended lifespan. *eLife*, *9*. https://doi.org/10.7554/elife.51089.
8. Ma, S., & Gladyshev, V. N. (2017). Molecular signatures of longevity: Insights from cross-species comparative studies. *Seminars in Cell & Developmental Biology*, *70*, 190–203. https://doi.org/10.1016/j.semcdb.2017.08.007.

Part IV.

Genetic Frontiers and Ethics

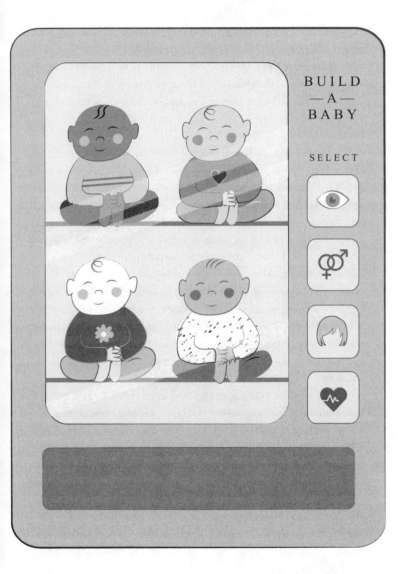

The 2020 Nobel Prize in Chemistry was awarded to Jennifer Doudna and Emmanuelle Charpentier for their role in developing CRISPR. As part of the announcement of the winners, Claes Gustafsson, chair of the Nobel Committee for Chemistry, warned, "The enormous power of this technology means that we need to use it with great care."[1] The power of biotechnology is often associated with risk. Many of the chapters in the previous parts of this anthology have touched on ethical issues associated with developments in biotechnology. They cannot be ignored and come to the forefront in part IV.

In 2019 He Jiankui shocked the world by announcing that he had edited the genomes of two young babies with traits that could be passed on to the next generation. Similar news stories about "CRISPR babies," "artificial wombs," "chimeric humans," and biohackers live-streaming their attempts at gene editing on Facebook have kicked off fierce debates on the ethics of "playing God" and the safety of trying an experimental technique on human subjects. These examples only hint at the many moral questions that emerging technologies pose—not only for medicine but for many other applications of biotechnology. Our regulatory frameworks are being outmoded by the rapid scientific expansion we are experiencing. At the same time, the introduction of new techniques is outpacing our understanding of what they do, or could do.

Putting rules in place to regulate biotechnology is a difficult feat. Countries have to consider safety, ethics, and competitiveness. If the regulations are too strict, countries will lose their competitive edge when researchers move abroad or do their experiments in secret, as He Jiankui did. This is the worst possible outcome.

What then are the roles of funders, journal editors, tenure committees, regulators, and philosophers in setting new ethical standards and regulations, and who is responsible for their oversight? The thought-provoking articles of part IV help inform and frame the debates. Their place at the end of the book reflects their ultimate importance.

Note

1. Rose, J. (2000, October 7). "This has revolutionised the molecular life sciences." [Video interview]. The Nobel Prize. [Website]. Prize announcement. https://www.nobelprize.org/prizes/chemistry/2020/prize-announcement/. Quotation spoken at 14:40.

Chapter 27

Scared of CRISPR? 45 Years on, IVF Shows How Fears of New Medical Technology Can Fade

PATRICIA A. STAPLETON

THE FIRST "TEST-TUBE BABY" made headlines around the world in 1978, setting off intense debate on the ethics of researching human embryos and reproductive technologies. Every breakthrough since then has raised the same questions about "designer babies" and "playing God," but public response has grown more subdued rather than more engaged as assisted reproductive technologies have become increasingly sophisticated and powerful.

As the science has advanced, doctors are able to perform more complex procedures with better-than-ever success rates.[1] This progress has made in vitro fertilization (IVF) and associated assisted reproductive technologies relatively commonplace. Over one million babies have been born in the United States using IVF since 1985. And Americans' acceptance of these technologies has grown along with their increased usage as we've gotten used to the idea of physicians manipulating embryos.

But the ethical challenges posed by these procedures remain—and in fact are increasing along with our gene-editing capabilities. While still a long way from clinical use, completed experiments by scientists in Oregon,[2] as well as in China, have demonstrated that genes in a human embryo can be edited, bringing us one step closer to changing the DNA that we pass along to our descendants. As the state of the science continues to advance, ethical issues must be addressed before the next big breakthrough.

Birth of the Test-Tube Baby Era

Louise Brown was born in the United Kingdom on July 25, 1978. Known as the first test-tube baby, she was a product of IVF, a process where an egg is fertilized by sperm outside a woman's body before being implanted in the womb. IVF opened up the possibility for infertile parents to have their own biologically related children. But Brown's family was also subjected to vicious hate mail, and groups opposed to IVF warned that it would be used for eugenic experiments, leading to a dystopian future where all babies would be genetically engineered.

The reaction in the United States had another layer to it compared with the reception in other developed countries. Here, research on embryos has historically been linked to the debate on abortion.[3] The 1973 Supreme Court decision to make abortion legal in *Roe v. Wade* fueled anti-abortion groups, which also oppose research on human embryos. Embryonic research and procedures offer the hope of eliminating devastating diseases, but scientists destroy embryos in the process.

Under pressure from these groups over the ethical implications of embryo creation and destruction, Congress issued a moratorium in 1974 on federally funded clinical research on embryos and embryonic tissue, including IVF, infertility, and prenatal diagnosis. To this day, federal funds are still not available for this type of work.

In hindsight, the intense media attention and sharp negative response from anti-abortion groups to IVF didn't accurately represent overall public opinion. The majority of Americans (60 percent) were in favor of IVF when polled in August 1978, and 53 percent of those polled said they would be willing to try IVF if they were unable to have a child. So, while the media coverage at the time helped inform the public of this new development, insensitively labeling Louise Brown a test-tube baby and warning of dystopian results didn't stop Americans from forming positive opinions of IVF.

Is Embryonic Research a Moral Issue?

In the years since IVF was introduced for use in humans, scientists have developed several new technologies—from freezing eggs to genetically testing embryos before implantation—that have improved the patient experience as well as the chances that IVF will result in the birth of a baby.[4] The announcement of each of these breakthroughs resulted in flurries of media attention to the ethical challenges raised by this type of research, but there has been no consensus—social, political, or scientific—on how to proceed.

Americans' general opinion on assisted reproductive technologies has remained positive. Despite opposition groups' efforts, surveys show that Americans have separated the issue of abortion from embryonic research. A Pew Research Center poll from 2013 revealed that only 12 percent of Americans said they personally considered using IVF to be morally wrong. That's a significant decrease from the 28 percent of respondents in 1978 who had said they opposed the procedure because it was "not natural." In addition, the 2013 poll showed that twice as many Americans (46 percent) said they did not consider using IVF to be a moral issue compared with those (23 percent) who said they did not consider having an abortion to be a moral issue.[5]

Why We Need to Pay Attention

Although most Americans don't think of embryonic research and procedures like IVF as a moral issue or as morally wrong, the introduction of new technologies is outpacing Americans' understanding of what they actually do. Polls from 2007 to 2008 showed that only 17 percent of respondents said they were "very familiar" with stem cell research, and the polls found a "relative absence of knowledge about even the most prominent of the embryo-research issues" (p. 49).[6] When Americans are asked more specific questions that explain IVF, they show less support for certain procedures, like freezing and storing eggs or using embryos for scientific research. Despite ongoing scientific experimentation with CRISPR, surveys show that nearly 69 percent of Americans have not heard or read much or know nothing at all about gene editing.[7] Additionally, support for gene editing depends on how the technology is to be used.

A majority of Americans generally accept gene editing if the purpose is to improve the health of a person or prevent a child from inheriting a debilitating disease. Some successful experiments have used a gene-editing technique that allowed researchers to correct a genetic defect in human embryos that causes heart disease.[8] This type of progress falls into the category that most Americans would support.

But the technique that's used to make this correction, known as CRISPR-Cas9, can potentially be used for editing genes in other ways, not just to eliminate diseases. Ongoing experimentation and new advances open the door to many possibilities in gene editing, including ones unrelated to health, such as changes to appearance or other physical characteristics.

Advancements in assisted reproductive technologies have happened rapidly over the last few decades, leading to over five million births worldwide. But as common as these procedures have become, scientists are not yet in agreement over how to integrate CRISPR and gene editing into the IVF tool kit. There are concerns about changing the genomes of human embryos destined to be babies, particularly since any modifications would be passed on to future generations. Scientific committees have noted that decisions on whether and how to use gene editing should be revisited on a regular basis. The newest breakthroughs with CRISPR are providing us with such opportunities.

We should focus our attention on answering the ethical questions that have long gone unanswered: What are the boundaries to this type of research? Who decides what is an ethical use of CRISPR? What responsibility do we have to people affected by genetic conditions? Who pays for these medical procedures? How will this research and potential clinical use be regulated?

The successful use of assisted reproductive technologies has skyrocketed in the last decade, making Americans complacent about some of the ethical concerns that these procedures raise. It's important that we engage with these issues now, before gene editing becomes as familiar to us as IVF.

Notes

1. Wade, J. J., MacLachlan, V., & Kovacs, G. (2015). The success rate of IVF has significantly improved over the last decade. *Australian and New Zealand Journal of Obstetrics and Gynaecology, 55*(5), 473–476. https://doi.org/10.1111/ajo.12356.
2. Ma, H., Marti-Gutierrez, N., Park, S.-W., Wu, J., Lee, Y., Suzuki, K., Koski, A., Ji, D., Hayama, T., Ahmed, R., Darby, H., Van Dyken, C., Li, Y., Kang, E., Park, A.-R., Kim, D., Kim, S.-T., Gong, J., Gu, Y., . . . Mitalipov, S. (2017). Correction of a pathogenic gene mutation in human embryos. *Nature, 548*(7668), 413–419. https://doi.org/10.1038/nature23305.
3. Wertz, D. C. (2002). Embryo and stem cell research in the United States: History and politics. *Gene Therapy, 9*(11), 674–678. https://doi.org/10.1038/sj.gt.3301744.
4. Moustafa Kamel, R. (2013). Assisted reproductive technology after the birth of Louise Brown. *Journal of Reproduction & Infertility, 14*(3). https://doi.org/10.4172/2161-0932.1000156.
5. *Abortion viewed in moral terms: Fewer see stem cell research and IVF as moral issues.* (2013, August 15). Pew Research Center. https://www.pewresearch.org/religion/2013/08/15/abortion-viewed-in-moral-terms/.

6. Levin, Y. (2008). Public opinion and the embryo debates. *New Atlantis* (Spring). https://thenewatlantis.com/wp-content/uploads/legacy -pdfs/20080607_TNA20Levin.pdf.

7. Blendon, R. J., Gorski, M. T., & Benson, J. M. (2016). The public and the gene-editing revolution. *New England Journal of Medicine, 374*(15), 1406–1411. https://doi.org/10.1056/nejmp1602010.

8. Ma, H., Marti-Gutierrez, N., Park, S.-W., Wu, J., Lee, Y., Suzuki, K., Koski, A., Ji, D., Hayama, T., Ahmed, R., Darby, H., Van Dyken, C., Li, Y., Kang, E., Park, A.-R., Kim, D., Kim, S.-T., Gong, J., Gu, Y., . . . Mitalipov, S. (2017). Correction of a pathogenic gene mutation in human embryos. *Nature, 548*(7668), 413–419. https://doi.org/10.1038/nature23305.

Chapter 28

How Can a Baby Have 3 Parents?

JENNIFER BARFIELD

IT SEEMS IMPOSSIBLE, RIGHT? We are taught that babies are made when a sperm and an egg come together, and the DNA from these two cells combine to make a unique individual, with half the DNA from the mother (one parent) and half from the father (the other parent). So how can a third person be involved in this process?

To understand the idea of three-parent babies, we have to talk about DNA. Most people are familiar with the double helix–style DNA, which makes up the 23 pairs of chromosomes and is found in the nucleus of every cell in our body. It provides the instructions for building an entire organism and for assembling the proteins that drive our existence from conception to death. The DNA in the nucleus is not, however, the only kind of DNA required for us to exist. There is also DNA tucked away in little compartments called mitochondria that are found inside all of the cells in your body.

Remember the mitochondria? Reach back to memories of middle or high school science class. A mitochondrion is that bean-shaped organelle often drawn with a squiggly line filling it in and is called the powerhouse of the cell. Each cell in the body, including eggs and sperm, requires energy to carry out all of its functions. Cells without functioning mitochondrial DNA (mtDNA) are like cars without gas.

Unlike nuclear DNA, mtDNA is not created by the combination of male and female DNA. In most cases, mitochondria are inherited only from your mother, meaning that the ones in the fertilized egg are the ones that will be replicated in every cell of your body during your development and for the rest of your life. Just like nuclear DNA, mtDNA can have mutations that lead to serious, debilitating diseases and, in some cases, to infertility for a woman carrying the defective mitochondria.

Enter the third parent.

The Third Parent

In 2016 a baby was born to a couple who had struggled with the consequences of mtDNA mutations that cause Leigh syndrome, a progressive neurometabolic disorder.[1] When defective mitochondria of the woman's egg were replaced with mitochondria from a donor who did not carry the mutation, the resulting child carried DNA from three people: the female nuclear DNA donor, the male nuclear DNA, and the female mtDNA donor. This was the first baby born using this technique.

This technique, termed *mitochondrial replacement*, can be thought of like an organ transplant or, rather, an organelle transplant. There are some significant differences, however, that have caused concern among legislators, resulting in a ban on mitochondrial replacement in the United States in 2015.[2]

Unlike an organ transplant, the effects of mitochondrial replacement will persist in future generations of offspring. Also, the replacement will affect every tissue in the body, rather than just one body system, such as would happen to the cardiovascular system after a heart transplant.

Even so, these donated mitochondria are naturally occurring and already persisting in our population. They are not genetically engineered or altered in any way. Thus, as long as they are functioning properly, there is no demonstrated risk to the offspring from a health standpoint beyond the

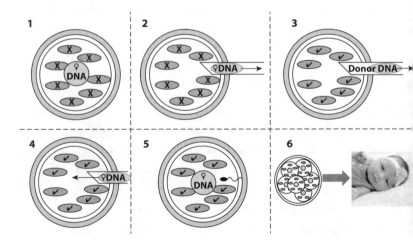

naturally occurring risks of spontaneous mutations, though this is a point of debate.[3]

Since 2016, it's difficult to say how many of these three-parent procedures have been done and how many resulted in successful pregnancies, but many countries are now exploring whether and how to use this technology. The ban in the United States has halted its use here, but other countries have made different decisions; the United Kingdom, for instance, has approved it.

Is the Mitochondrial Donor a Parent?

So how much of a parent is a woman who donates her mitochondria?

The short answer is not much. More than 99 percent of the proteins in your body are encoded by the DNA in the nucleus of your cells. Traits such as hair color, eye color, and height, for example, are all encoded by nuclear DNA, while

How to make a three-parent baby. (1) The egg from the mother contains the haploid DNA in the nucleus (circle) and faulty mitochondria (ovals with Xs). (2) The nucleus with its DNA is removed from the mother's egg using a tiny pipette. (3) The nucleus in the mitochondrial donor egg is removed, leaving behind the healthy mitochondria (ovals with checkmarks). (4) The DNA from the mother is transferred to the donor egg with its healthy mitochondria. (5) The result is an egg with nuclear DNA from the mother and mitochondrial DNA from the mitochondria donor; this egg can then be fertilized with the father's sperm. (6) As cells replicate during embryo development, each new cell will have the combined mother and father's diploid DNA in its nucleus and the donor's replicated mitochondria and mitochondrial DNA. Note that fertilization can occur before or after the transfer of nuclear DNA to the donor egg. If it happens before, then both the mother's and the father's DNA will be transferred to the mitochondrial donor egg after the donor's DNA is removed. If it occurs after, as diagrammed here, then the egg will be fertilized after the mother's DNA is transferred to the donor egg. *Jennifer Barfield, CC BY-ND* genes written on mtDNA are primarily related to energy production and metabolism.[4]

Thus a three-parent baby will resemble the man and woman whose sperm and egg combined to produce the 23 chromosomes in the nucleus of that first cell. It's important for people to understand these distinctions, as headlines announcing the birth of three-parent babies will likely continue to appear. Speculation over what it means could run wild without an understanding of the underlying science.

One thing is certain: for women who struggle with infertility caused by a mutation in their mitochondrial DNA, or who have the potential to pass on a significant mitochondrial genetic defect, this new technique provides hope that they may one day be able to have a healthy child that genetically represents them and their partner—with a little help from a third party.

Notes

1. Zhang, J., Liu, H., Luo, S., Lu, Z., Chávez-Badiola, A., Liu, Z., Yang, M., Merhi, Z., Silber, S. J., Munné, S., Konstantinidis, M., Wells, D., Tang, J. J., & Huang, T. (2017). Live birth derived from oocyte spindle transfer to prevent mitochondrial disease. *Reproductive BioMedicine Online, 34*(4), 361–368. https://doi.org/10.1016/j.rbmo.2017.01.013.
2. Adashi, E. Y., & Cohen, I. G. (2017). Mitochondrial replacement therapy: Unmade in the USA. *JAMA, 317*(6), 574. https://doi.org/10.1001/jama.2016.20935.
3. Hamilton, G. (2015). The hidden risks for "three-person: babies. *Nature, 525*(7570), 444–446. https://doi.org/10.1038/525444a.
4. Taanman, J.-W. (1999). The mitochondrial genome: Structure, transcription, translation and replication. *Biochimica et Biophysica Acta (BBA)—Bioenergetics, 1410*(2), 103–123. https://doi.org/10.1016/s0005-2728(98)00161-3.

Chapter 29

Ethicists Need More Flexible Tools for Evaluating Gene-Edited Food

CHRISTOPHER J. PRESTON and TRINE ANTONSEN

IS THERE NOW A WAY TO GENETICALLY engineer crops to create food that people can confidently consider to be natural?

Gene-editing technology sounds like it might offer this possibility. Because genome editing can alter an organism's genetic material, or genome, without introducing genes from other species, its advocates argue that it can sidestep most of the ethical and regulatory worries plaguing organisms with added "transgenes," which are genes from other species.[1] Some even contend these "cisgenic" products are natural enough to count as organic.

Genome editing, its boosters say, can make changes that look almost evolutionary. Arguably, these changes could have happened by themselves through the natural course of events, if anyone had the patience to wait for them. Conventional breeding for potatoes resistant to late blight is theoretically possible, for example, but it would take a lot of time.

Speaking as ethicists who study how technology alters human–nature relations, we can understand why advocates see the ethics of genome editing this way. If "crossing species lines" is the measure of whether a technique counts as "natural" or not, then genome editing appears to pass a naturalness test. Although we understand the advantages of speeding up species adaptation, we don't think an ethics hinging on the idea of "cisgenesis" is adequate. We propose a better ethical lens to apply in its place.

Naturalness and Species Lines

Our work is part of a four-year project funded by the Norwegian Research Council that is studying how gene editing could change how we think about food. The work brings together researchers from universities and scientific institutes in Norway, the United Kingdom, and the United States to compare a range of techniques for producing useful new crops.

Our project is not focused on the safety of the crops under development, something that obviously requires concerted scientific investigation of its own. Although the safety of humans and the health of the environment is ethically crucial when developing new foods, other ethical issues must also be considered. To see this, consider how objections to genetically modified organisms go far beyond safety. Ethical issues in food sovereignty range broadly across farmer choice, excess corporate power, economic security, and other concerns. Ethical acceptability requires clearing a much higher bar than safety alone.

Although we believe gene editing may have promise for addressing the agricultural challenges caused by rising global population, climate change, and the overuse of chemical pesticides, we don't think an ethical analysis based entirely on "crossing species lines" and "naturalness" is adequate. It is already clear that declaring gene-edited food ethical for this reason has not satisfied all of gene editing's critics. As Ricarda Steinbrecher, a molecular biologist who's cautious about gene editing, has written, "Whether or not the DNA sequences come from closely related species is irrelevant, the process of genetic engineering is the same, involving the same risks and unpredictabilities, as with transgenesis" (p. 5).[2]

Comments of this kind suggest that talking about species lines is an unreliable guide. Species and subspecies boundaries are notoriously infirm. Charles Darwin himself conceded the point: "I look at the term species, as one arbitrarily given for the sake of convenience to a set of individuals closely resembling each other" (p. 34).[3] The 2005 edition of *Mammal Species of the World* demonstrated this arbitrariness by collapsing all 12 subspecies of American cougars into a single species: *Puma concolor*. In 2017 the Cat Classification Task Force revised family Felidae again.

If species lines are not clearly drawn, then claiming naturalness for not having crossed them is, in our view, a shaky guide. The lack of clarity matters because a premature ethical green light could mean a premature regulatory green light, with broad implications for both agricultural producers and consumers.

The Integrity Lens

We think a more reliable ethical measure is to ask how a technique for crop breeding interferes with the integrity of the organism being altered. The term *integrity* already has application in environmental ethics, ecology,[4] cell biology,[5] interhuman ethics, organic agriculture,[6] and genetics.[7] A unifying theme in all these domains is that integrity points toward some kind of functional wholeness for an organism, a cell, a genome, or an ecosystem. The idea of maintaining integrity tracks a central intuition about being cautious before interfering too much with living systems and their components.

The integrity lens makes it clear why the ethics of gene editing may not be radically different from the ethics of genetic modification using transgenes. The cell wall is still penetrated by the gene-editing components. The genome of the organism is cut at a site chosen by the scientist, and a repair is initiated that (it is hoped) will result in a desired change to the organism. When it comes to the techniques involved with gene-editing a crop or other food for a desired trait, integrity is compromised at several levels, and none has anything to do with crossing species lines. The integrity lens makes it clear the ethics is not resolved by debating natural-ness or species boundaries.

Negotiation of each other's integrity is a necessary part of human-to-human relations. If adopted as an ethical practice in the field of biotechnology, it might provide a better guide in attempts to accommodate different ethical, ecologi-cal, and cultural priorities in policy making.[8] An ethic with a central place for discussing integrity promises a framework that is both more flexible and discerning.[9]

As new breeding techniques create new ethical debates over food, we think the ethical toolbox needs restocking. Talking about crossing species lines simply isn't enough. If Darwin had known about gene editing, we think he would have agreed.

Notes

1. Carroll, D., Van Eenennaam, A. L., Taylor, J. F., Seger, J., & Voytas, D. F. (2016). Regulate genome-edited products, not genome editing itself. *Nature Biotechnology, 34*(5), 477–479. https://doi.org/10.1038/nbt .3566.
2. Steinbrecher, R. A. (2015, December). *Genetic engineering in plants and the "new breeding techniques (NBTs)": Inherent risks and the need to regulate.* EcoNexus. https://www.econexus.info/files /NBT%20Briefing%20-%20EcoNexus%20December%202015.pdf.
3. Darwin, C. (1859/2007). *On the origin of species.* Cosimo.
4. Wurtzebach, Z., & Schultz, C. (2016). Measuring ecological integrity: History, practical applications, and research opportunities. *BioScience, 66*(6), 446–457. https://doi.org/10.1093/biosci/biw037.
5. Rodicio, R., & Heinisch, J. J. (2010). Together we are strong—cell wall integrity sensors in yeasts. *Yeast, 27*(8), 531–540. https://doi.org/10 .1002/yea.1785.
6. Bueren, E. T., & Struik, P. C. (2005). Integrity and rights of plants: Ethical notions in organic plant breeding and propagation. *Journal of Agricultural and Environmental Ethics, 18*(5), 479–493. https://doi .org/10.1007/s10806-005-0903-0.
7. Waters, R. (2006). Maintaining genome integrity. *EMBO Reports, 7*(4), 377–381. https://doi.org/10.1038/sj.embor.7400659.
8. Stephan, H. R. (2012). Revisiting the transatlantic divergence over GMOs: Toward a cultural-political analysis. *Global Environmental Politics, 12*(4), 104–124. https://doi.org/10.1162/glep_a_00142.
9. Preston, C., & Antonsen, T. (2021). Integrity and agency: Negotiating new forms of human-nature relations in biotechnology. *Environmental Ethics, 43*(1), 21–41. https://doi.org/10.5840/enviroethics202143020.

Chapter 30

Lab-Grown Embryos and Human-Monkey Hybrids: Medical Marvels or Ethical Missteps?

SAHOTRA SARKAR

IN ALDOUS HUXLEY'S 1932 novel *Brave New World*, people aren't born from a mother's womb. Instead, embryos are grown in artificial wombs until they are brought into the world, a process called ectogenesis. In the novel, technicians in charge of the hatcheries manipulate the nutrients they give the fetuses to make the newborns fit the desires of society. Two important scientific developments suggest that Huxley's imagined world of functionally manufactured people is no longer far-fetched.

On March 17, 2021, an Israeli team announced that it had grown mouse embryos for 11 days—about half of the complete gestation period—in artificial wombs that were essentially bottles.[1] Until this experiment, no one had grown a mammal embryo outside a womb this far into pregnancy. Then, on April 15, 2021, a US and Chinese team announced that it had grown, for the first time, embryos that included both human and monkey cells to a stage where organs began to form.[2]

As both a philosopher and a biologist, I cannot help but ask how far researchers should take this work. While creating chimeras—the name for creatures that are a mix of organisms—might seem like the more ethically fraught of these two advances, ethicists think the medical benefits far outweigh the ethical risks. Ectogenesis has not been put under nearly as much scrutiny as chimeras have; however, it could have far-reaching impacts on individuals and society.

Growing Life in an Artificial Womb

When in vitro fertilization (IVF) first emerged in the late 1970s, the press called IVF embryos "test-tube babies," even though they are nothing of the sort. These embryos are implanted in the uterus within a day or two after doctors fertilize an egg in a petri dish.

Before the Israeli experiment, researchers had not been able to grow mouse embryos outside the womb for more than four days because providing the embryos with enough oxygen had been too hard to do. The Israelis spent seven years creating a system of slowly spinning glass bottles and con- trolled atmospheric pressure that simulates the placenta and provides oxygen. This development is a major step toward ectogenesis, and scientists expect that it will be possible to extend mouse development farther, possibly to full term outside the womb. This will likely require new techniques, but at this point it is a problem of scale—being able to accommodate a larger fetus. This appears to be a simpler challenge to overcome than figuring out something totally new like supporting organ formation.

The Israeli team plans to deploy its techniques with human embryos. Since mice and humans have similar developmental processes, it is likely that the team will succeed in growing human embryos in artificial wombs. To do so, though, members of the team need permission from their ethics board.

CRISPR—a technology that can cut and paste genes—already allows scientists to manipulate an embryo's genes after fertilization. Once fetuses can be grown outside the womb, as in Huxley's world, researchers will also be able to modify their growing environments to influence what physical and behavioral qualities these parentless babies exhibit.[3] Science still has a way to go before fetus development and births outside a uterus become a reality, but researchers are getting closer. The question now is how far humanity should go down this path.

Human–Monkey Hybrids

Human–monkey hybrids might seem to be a much scarier prospect than babies born from artificial wombs. But, in fact, this research is more of a step toward an important medical development than one toward an ethical minefield. If scientists can grow human cells in monkeys or other animals, it should be possible to grow human organs too.[4] This would solve the problem of organ shortages around the world for people needing transplants.

But keeping human cells alive in the embryos of other animals for any length of time has proved to be extremely difficult. In the human-monkey chimera experiment, a team of researchers implanted 25 human stem cells into embryos of crab-eating macaques, a type of monkey. The researchers then grew these embryos for 20 days in petri dishes. After 15 days, the human stem cells had disappeared from most of the embryos. But at the end of the 20-day experiment, three embryos still contained human cells that had grown as part of the region of the embryo where they were embedded. For scientists, the challenge now is to figure out how to maintain human cells in chimeric embryos for longer.

Regulating These Technologies

Some ethicists have begun to worry that researchers are rushing into a future of chimeras without adequate preparation. Their main concern is the ethical status of chimeras that contain human and nonhuman cells—especially if the human cells integrate into sensitive regions such as a monkey's brain. What rights would such creatures have?

There seems, though, to be an emerging consensus that the potential medical benefits justify a step-by-step extension of this research. Many ethicists are urging public discussion of appropriate regulation to determine how close to viability these embryos should be grown. One proposed solution is to limit their growth to the first trimester of pregnancy, by which time most organ formation will have taken place. Given that researchers don't plan to grow these embryos beyond the stage when they can harvest rudimentary organs, I don't believe chimeras are ethically problematic compared with the true test-tube babies of Huxley's world.

Few ethicists have broached the problems posed by a hypothetical ability to use ectogenesis to engineer human beings that fit societal desires. Researchers have yet to conduct experiments on human ectogenesis, and for now, scientists lack the techniques to bring the embryos to full term. However, without regulation, I believe researchers are likely to try these techniques on human embryos—just as the now-infamous He Jiankui used CRISPR to edit human babies without properly assessing safety and desirability. Techno-logically, it is a matter of time before mammal embryos can be brought to term outside the body.

While people may be uncomfortable with ectogenesis today, this discomfort could pass into familiarity, as happened with IVF. But scientists and regulators would do well to reflect on the wisdom of permitting a process that could allow someone to engineer human beings without parents. As critics have warned, in the context of CRISPR-based genetic enhancement, pressure to change future generations to meet societal desires will be unavoidable and dangerous, regardless of whether that pressure comes from an authoritative state or cultural expectations.[5] In Huxley's imagination, hatcheries run by the state grew large numbers of identical individuals as needed. That would be a very different world from ours today.

Notes

1. Aguilera-Castrejon, A., Oldak, B., Shani, T., Ghanem, N., Itzkovich, C., Slomovich, S., Tarazi, S., Bayerl, J., Chugaeva, V., Ayyash, M., Ashouokhi, S., Sheban, D., Livnat, N., Lasman, L., Viukov, S., Zerbib, M., Addadi, Y., Rais, Y., Cheng, S., . . . Hanna, J. H. (2021). Ex utero mouse embryogenesis from pre-gastrulation to late organogenesis. *Nature*, *593*(7857), 119–124. https://doi.org/10.1038/s41586-021-03416-3.

2. Tan, T., Wu, J., Si, C., Dai, S., Zhang, Y., Sun, N., Zhang, E., Shao, H., Si, W., Yang, P., Wang, H., Chen, Z., Zhu, R., Kang, Y., Hernandez-Benitez, R., Martinez Martinez, L., Nuñez Delicado, E., Berggren, W. T., Schwarz, M., . . . Izpisua Belmonte, J. C. (2021). Chimeric contribution of human extended pluripotent stem cells to monkey embryos ex vivo. *Cell*, *184*(8), 2020–2032.e14. https://doi.org/10.1016/j.cell .2021.03.020.

3. Wu, G., Bazer, F. W., Cudd, T. A., Meininger, C. J., & Spencer, T. E. (2004). Maternal nutrition and fetal development. *Journal of Nutrition*, *134*(9), 2169–2172. https://doi.org/10.1093/jn/134.9.2169.

4. Greely, H. T., & Farahany, N. A. (2021). Advancing the ethical dialogue about monkey/human chimeric embryos. *Cell*, *184*(8), 1962–1963. https://doi.org/10.1016/j.cell.2021.03.044.

5. Sparrow, R. (2011). A not-so-new eugenics. *Hastings Center Report*, *41*(1), 32–42. https://doi.org/10.1002/j.1552-146x.2011.tb00098.x.

Chapter 31

Those Designer Babies Everyone Is Freaking Out about Are Not Likely to Happen

A. CECILE J. W. JANSSENS*

WHEN ADAM NASH WAS STILL AN EMBRYO, living in a dish in the lab, scientists tested his DNA to make sure it was free of Fanconi anemia, the rare inherited blood disease from which his sister, Molly, suffered.[1] They also checked his DNA for a marker that would reveal whether he shared the same tissue type. Molly needed a donor match for stem cell therapy, and

---

* *Editor's note*: Sadly, the author of this article, A. Cecile J. W. Janssens, passed away in September 2022.

her parents were determined to find one. Adam was conceived so that the stem cells in his umbilical cord could be the lifesaving treatment for his sister.

Adam Nash is considered to be the first designer baby, born in 2000 after in vitro fertilization with pre-implantation genetic diagnosis, a technique used to choose desired characteristics. The media covered the story with empathy for the parents' motives but not without reminding the reader that "eye color, athletic ability, beauty, intelligence, height, stopping a propensity towards obesity, guaranteeing freedom from certain mental and physical illnesses, all of these could in the future be available to parents deciding to have a designer baby."[2]

Designer babies have thus been called the future-we-should-not-want, prognosticated with each new reproductive technology or intervention. But the babies never came and are nowhere close to coming. I am not surprised.

I study the prediction of complex diseases and human traits that result from interactions among multiple genes and lifestyle factors. This research shows that geneticists cannot read the genetic code like tea leaves to foretell who will be above average in intelligence or athleticism. Such traits and diseases that result from multiple genes and lifestyle factors cannot be predicted from DNA alone and cannot be designed.[3] Not now. And very likely not ever.

Proclaiming Designer Babies Are Next

The inevitable rise of designer babies was proclaimed in 1978—after the birth of Louise Brown, the first baby resulting from in vitro fertilization (IVF)—to be the next step toward "a brave world where parents can select their child's gender and traits."[4] The same situation occurred in 1994 when a 59-year-old British woman stretched the limits of nature by giving birth to twins using donated eggs that were implanted in her womb at a fertility clinic in Italy. The response was the same in 1999 when a fertility clinic in Fairfax, Virginia, offered sex selection of embryos to screen against diseases that happen only in boys.

The 21st century brought new developments. In 2013, when 23andMe was granted a patent for a tool that predicts the likelihood of traits in babies based on the DNA of two parents, the question of patenting designer babies was raised. In 2016, when the United Kingdom permitted a woman to donate her healthy mitochondria to a couple who were using IVF to conceive a child, thereby raising the number of parents to three, fears of unnatural children rose again. In 2018, when Genomic Prediction, a New Jersey company, announced that its DNA screening panel for embryos would also assess the risk for complex diseases such as type 2 diabetes and heart disease, which are caused by multiple genes, fears of engineering babies with high intelligence or athletic prowess were roused. The same issues arose again later that year when He Jiankui reported, at the Second International Summit on Human Genome Editing in Hong Kong, that he had successfully edited the DNA of twin baby girls.

The arrival of designer babies, portended and dreaded, has not come to pass with technology. It's been the same story for decades—with the same "desirable" traits and the same assumption that parents would want to select these traits if technology allowed for it. But no one seems to be questioning whether these traits are solely a product of our genes such that they could be selected or edited in embryos. Wondering and worrying about designer babies was understandable in the early days, but the repetition of these supposed fears now suggests a general lack of understanding about how DNA, and the genes it encodes, works.

Designing Favorable Traits in Babies Is Not Simple

Although there are exceptions, DNA generally differs between people in two ways: there are DNA mutations and DNA variations.

Mutations cause rare diseases such as Huntington's disease and cystic fibrosis, which are caused by a single gene. Mutations in the *BRCA* genes substantially increase women's risk for breast and ovarian cancer. Selecting embryos that do not have these mutations removes a major cause of these cancers, yet women who don't have *BRCA* mutations can still develop breast or ovarian cancer through other causes.

Variations are changes in the genetic code that are more widespread than mutations and are associated with common traits and diseases. Having a DNA variant may increase the likelihood that a person will possess a trait or develop a disease, but the variant does not determine or cause it. Association means that, in several studies of large populations, a DNA variant was more frequently found in people with the trait than in those without it, although often only slightly more frequently. These variants don't determine a trait but increase its likelihood of expression by interacting with other DNA variants and nongenetic influences such as upbringing, lifestyle, and other features of a person's environment. To design such traits in embryos would require multiple DNA changes in multiple genes, as well as orchestrating relevant environmental influences too.

Let's compare DNA to driving a car. DNA mutations are like a flat tire or failing brakes: technical problems that make driving problematic, no matter where you drive. DNA variations are like the model of car, which may affect the driving experience and even create problems over time. For example, a convertible is a delight when cruising down Hollywood's Sunset Boulevard on a sunny afternoon in summer, but a roofless ride would be a misery when crossing a mountain pass in a winter snowstorm. Whether features of the car are an asset or a liability depends on the context, and that context might change—they are never ideal all the time and in every circumstance.

Another Hurdle

Most DNA mutations do nothing else other than cause the disease, but DNA variations may play a role in many diseases and traits. Take variations in the *MC1R* "red hair" gene, which not only increases the chance a child will have red hair but also increases the child's risk of having skin cancer later in life. Or take variations in the *OCA2* and *HERC2* "eye color" genes, which are also associated with the risk of various cancers, Parkinson's, and Alzheimer's. To be sure, these are statistical associations found in population studies—some may be confirmed by further researcher; others may not. But the message is clear: editing DNA variations for "desirable" traits may have adverse consequences, including many that scientists don't know about yet.

We can consider this possibility for He Jiankui's gene-edited babies. By trying to make the babies resistant to HIV infection, the aim of his research, He Jiankui might have increased their susceptibility to infections by West Nile virus or influenza.[5]

To be sure, even though complex traits such as intelligence, athleticism, and musicality cannot be selected or designed, there will be opportunists who will try to offer these traits, even if prematurely and unsupported by science. Consider what Stephen Hsu, the cofounder of Genomic Prediction, said about testing embryos for polygenic risk, which is the risk of developing a disease based on multiple genes: "I think people are going to demand that. If we don't do it, some other company will."[6] And what researcher He said: "There will be someone, somewhere, who is doing this. If it's not me, it's someone else."[7] People need to be protected against such irresponsible and unethical use of DNA testing and editing.

Science brought about incredible progress in reproductive technology but didn't bring designer babies one step closer to realization. The creation of designer babies is not limited by technology but by biology: the origins of common traits and diseases are too complex and intertwined for us to modify the DNA without introducing unwanted effects.

Notes

1. Verlinsky, Y., Rechitsky, S., Schoolcraft, W., Strom, C., & Kuliev, A. (n.d.). Designer babies—are they a reality yet? Case report: Simultaneous preimplantation genetic diagnosis for Fanconi anaemia and HLA typing for cord blood transplantation. *Reproductive BioMedicine Online, 1*(2), 77. https://doi.org/10.1016/S1472-6483(10)61943-8.
2. Oliver, M. (2000, October 4). Genetic parenting—"designer babies." *The Guardian.* https://www.theguardian.com/world/2000/oct/04/qanda.markoliver.
3. Janssens, A. C., & van Duijn, C. M. (2010). An epidemiological perspective on the future of direct-to-consumer personal genome testing. *Investigative Genetics, 1*(1). https://doi.org/10.1186/2041-2223-1-10.

4. Fleming, A. T. (1980, July 20). New frontiers in conception: Medical breakthroughs and moral dilemmas. *New York Times*.

5. Keynan, Y., Juno, J., Meyers, A., Ball, T. B., Kumar, A., Rubinstein, E., & Fowke, K. R. (2010). Chemokine receptor 5 Δ32 allele in patients with severe pandemic (H1N1) 2009. *Emerging Infectious Diseases, 16*(10), 1621–1622. https://doi.org/10.3201/eid1610.100108.

6. Quoted in Wilson, C. (2018, November 15). *Exclusive: A new test can predict IVF embryos' risk of having a low IQ*. NewScientist. https://www.newscientist.com/article/mg24032041-900-exclusive-a-new-test-can-predict-ivf-embryos-risk-of-having-a-low-iq/.

7. Quoted in Dangerous science in China. (2019, November 28). *Japan Times*. https://www.japantimes.co.jp/opinion/2018/11/29/editorials/dangerous-science-china/.

Chapter 32

Bioweapons Research Is Banned by an International Treaty—but Nobody Is Checking for Violations

GARY SAMORE

SCIENTISTS ARE MAKING DRAMATIC progress with techniques of "gene splicing" for modifying the genetic makeup of organisms. This work includes bioengineering pathogens for medical research, which can also be used to create deadly biological weapons. It's an overlap that's helped fuel speculation that the coronavirus SARS–CoV–2 was bioengineered at China's Wuhan Institute of Virology and that it subsequently "escaped" through a lab accident and caused the COVID–19 pandemic.

The world already has a legal foundation to prevent gene splicing for warfare: the 1972 Biological Weapons Convention.[1] Unfortunately, nations have been unable to agree on how to strengthen the treaty. Some countries have also pursued bioweapons research and stockpiling in violation of it.[2]

As a member of President Bill Clinton's National Security Council from 1996 to 2001, I had a firsthand view of the failure to strengthen the convention. From 2009 to 2013, as President Barack Obama's White House coordinator for weapons of mass destruction, I led a team that grappled with the challenges of regulating potentially dangerous biological research in the absence of strong international rules and regulations.

The history of the Biological Weapons Convention reveals the limits of international attempts at controlling the research and development of biological agents.

1960s–1970s: International Negotiations to Outlaw Biowarfare

The United Kingdom first proposed a global biological weapons ban in 1968.[3] Reasoning that bioweapons had no useful military or strategic purpose given the awesome power of nuclear weapons, the United Kingdom had ended its offensive bioweapons program in 1956. But the risk remained that other countries might consider developing bioweapons as a poor man's atomic bomb.[4] In the original British proposal, countries would have to identify facilities and activities with potential bioweapons applications. They would also need to accept on-site inspections by an international agency to verify that these facilities were being used for peaceful purposes.

Negotiations about the ban gained steam in 1969 when the Nixon administration ended America's offensive biological weapons program and supported the British proposal. In 1971 the Soviet Union announced its support—but only with the verification provisions stripped out. Since it was essential to get the Soviet Union on board, the United States and United Kingdom agreed to drop those requirements.

In 1972 the treaty was finalized. After gaining the required signatures, it took effect in 1975.

Under the convention, 183 nations have agreed not to "develop, produce, stockpile or otherwise acquire or retain" biological materials that could be used as weapons. They also agreed not to stockpile or develop any "means of delivery" for using them. The treaty allows "prophylactic, protective or other peaceful" research and development, including medical research. What the treaty lacks, however, is any mechanism for verifying that countries are complying with these obligations.

1990s: Revelations of Treaty Violations

This absence of verification was exposed as the convention's fundamental flaw two decades later, when it turned out that the Soviets had a great deal to hide.

In 1992 Russian president Boris Yeltsin revealed the Soviet Union's massive biological weapons program. Some of the program's reported experiments had involved making viruses and bacteria more lethal and resistant to treatment.[5] The Soviets also weaponized and mass-produced a number of dangerous naturally occurring viruses, including the anthrax and smallpox viruses, as well as the plague-causing bacterium *Yersinia pestis*. Yeltsin ordered the program's end in 1992 and the destruction of all its materials. But doubts remain over whether this order was fully carried out.

The same revolutionary techniques of gene splicing that scientists can use to breed climate change–resistant hybrid crops—like these experimental barley seeds created in a German lab—can also be used to create biological weapons. *Sean Gallup/Getty News Images via Getty Images Europe*

Another treaty violation came to light after the US defeat of Iraq in the 1991 Gulf War. United Nations inspectors discovered an Iraqi bioweapons stockpile, including 1,560 gallons (6,000 liters) of anthrax spores and 3,120 gallons (12,000 liters) of botulinum toxin. Both had been loaded into aerial bombs, rockets, and missile warheads, although Iraq never used these weapons.

In the mid-1990s, during South Africa's transition to majority rule, evidence emerged of the former apartheid regime's chemical and biological weapons program. As revealed by the South African Truth and Reconciliation Commission, the program focused on assassination. Techniques included infecting cigarettes with anthrax spores, sugar with salmonella, and chocolates with botulinum toxin.

In response to these revelations, as well as suspicions that North Korea, Iran, Libya, and Syria were also violating the treaty, the United States began urging other nations to close the verification gap. But despite 24 meetings over seven years, a specially formed group of international negotiators failed to reach agreement on how to do it.[6] The problems were both practical and political.

Monitoring Biological Agents

Several factors make it difficult to verify compliance with the bioweapons treaty.

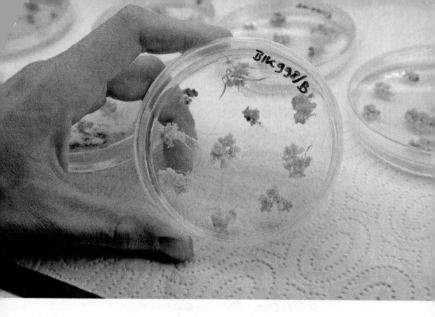

First of all, facilities for researching and producing biological agents for condoned purposes, such as vaccines, antibiotics, vitamins, biological pesticides, and certain foods, can produce biological weapons too. And some pathogens with legitimate medical and industrial uses can also be used for bioweapons.

Furthermore, large quantities of certain biological weapons can be produced quickly, by few personnel and in relatively small facilities. Hence, biological weapons programs are more difficult for international inspectors to detect than are nuclear or chemical programs, which typically require large facilities, numerous personnel, and years of operation.

So, an effective bioweapons verification process would require that enforcing nations identify a large number of civilian facilities. Inspectors would need to monitor the facilities regularly. The monitoring would need to be intrusive by allowing inspectors to demand "challenge inspections," meaning they could gain access on short notice to both known and suspected facilities.

Finally, developing bioweapons defenses, as permitted under the treaty, typically requires working with dangerous pathogens and toxins and even with delivery systems. So distinguishing legitimate biodefense programs from illegal bioweapons activities often comes down to intent—and intent is hard to verify.

Because of these inherent difficulties, verification faced stiff opposition.

Political Opposition to Bioweapons Verification

As the White House official responsible for coordinating the US negotiating position, I often heard concerns and objections from important government agencies. The Pentagon expressed fears that inspections of US biodefense installations would compromise national security or lead to false accusations of treaty violations. The Commerce Department opposed intrusive international inspections on behalf of the pharmaceutical and biotechnology industries. Such inspections might compromise trade secrets, department officials warned, or interfere with medical research or industrial production.

Germany and Japan, which also have large pharmaceutical and biotechnology industries, raised similar objections. China, Pakistan, Russia, and others opposed nearly all on-site inspections. Since the rules under which the negotiation group operated required consensus, any single country could block agreement.

In January 1998, seeking to break the deadlock, the Clinton administration proposed reduced verification requirements.

Nations could limit their declarations to facilities "especially suitable" for bioweapons uses, such as vaccine production facilities. Moreover, random or routine inspections of these facilities would instead be "voluntary" visits or limited challenge inspections—but only if approved by the executive council of a to-be-created international agency monitoring compliance with the bioweapons treaty.

But even this form of reduced verification failed to achieve consensus among the international negotiators. Finally, in July 2001, the George W. Bush administration rejected the Clinton proposal, ironically doing so on the grounds that it was not strong enough to detect cheating. With that, the negotiations collapsed.

Since then, nations have made no serious effort to establish a verification system for the Biological Weapons Convention. Even with the amazing advances scientists have made in genetic engineering since the 1970s, there are few signs that countries are interested in taking up the problem again. This is especially true in today's climate of accusations against China, which has refused to cooperate fully in determining the origins of the COVID-19 pandemic.

Notes

1. Gerstein, D., & Giordano, J. (2017). Rethinking the biological and toxin weapons convention? *Health Security*, *15*(6), 638–641. https://doi.org/10.1089/hs.2017.0082.
2. Zilinskas, R. A. (1997). Iraq's biological weapons. The past as future? *JAMA*, *278*(5), 418–424. https://doi.org/10.1001/jama.1997.03550050080037.
3. Sims, N. A. (2011). A simple treaty, a complex fulfillment: A short history of the biological weapons convention review conferences. *Bulletin of the Atomic Scientists*, *67*(3), 8–15. https://doi.org/10.1177/0096340211407400.
4. Horowitz, M. C., & Narang, N. (2013). Poor man's atomic bomb? Exploring the relationship between "weapons of mass destruction."

Journal of Conflict Resolution, *58*(3), 509–535. https://doi.org/10.1177/0022002713509049.

5. Hart, J. (2006). *Deadly cultures*. Harvard University Press. https://doi.org/10.4159/9780674045132-007. See chapter 6, "The Soviet Biological Weapons Program," pp. 132–156.

6. Butler, D. (2001). Bioweapons treaty in disarray as US blocks plans for verification. *Nature, 414*(6865), 675–675. https://doi.org/10.1038/414675a.

Chapter 33

From Coronavirus Tests to Open-Source Insulin and Beyond, "Biohackers" Are Showing the Power of DIY Science

ANDREW LAPWORTH

IN THE SPRING OF 2020, Will Canine and a team of amateur scientists in New York City developed an automated robot platform called Opentrons that dramatically hastened the turnaround time for COVID-19 test results from two weeks to 12 hours.[1] This innovative technology, which is now used around the world, helped save countless lives in the early days of the outbreak.

The technology's creators, affiliated with a "community lab for citizen scientists" called Genspace, are part of a growing international movement of so-called biohackers with roots stretching back 30 years or more. Biohacking, also known as do-it-yourself, or DIY, biology, takes cues from computer-hacking culture and uses the tools of biological science and biotechnology to carry out experiments and make tools outside any formal research institution.

But biohacking is under threat as governments, wary of potential risks, pass laws to restrict it. A more balanced approach is needed, for the benefit of science and society.

Who's Afraid of Biohacking?

As biohacking has gained increased visibility, it has also attracted increased scrutiny. Media coverage has played up the risks of biohacking, whether from malice ("bioterror") or by accident ("bioerror"). Local and national governments have also sought to legislate against the practice.

In August 2019, politicians in California introduced a law that forbids the use of CRISPR gene-editing kits outside professional labs. Australia has some of the world's most stringent regulations, with the Office of the Gene Technology Regulator monitoring the use of genetically modified organisms and risks to public health and safety. Some authorities have gone so far as to arrest biohackers on suspicion of bioterrorism.[2]

But such anxieties around biohacking are largely unfounded.

Ellen Jorgensen, cofounder of the Genspace community lab in New York City, argues that antagonistic responses overestimate the abilities of biohackers and underestimate their ethical standards. Research shows the great majority of biohackers (92 percent) work within community laboratories, many of which operate under the Ethical Code for Safe Amateur Bioscience drawn up by the community in 2011.[3]

Connoisseurs of Science

One way to think of biohackers is as "connoisseurs of science," which is what Belgian philosopher Isabelle Stengers calls them.[4] Somewhere between an expert and an amateur, a connoisseur can relate to scientific knowledge and practice in an informed way but can also pose new questions that scientists are unable to. Connoisseurs can hold scientists to account and challenge them when they skip over concerns. Connoisseurs highlight how science might be done better. Like other pursuits such as music or sport, science can benefit from a strong and vibrant culture of connoisseurs.

Biohackers are an important node in the relationship between science institutions and wider society. Stengers highlights how it is not enough for there to be a relationship between science and society. It is the nature and quality of this relationship that matters.

A Two-Way Relationship

Traditional models of science communication assume a one-way relationship between science and society at large, with scientists transmitting knowledge to a public that passively receives it. Biohackers engage people, instead, as active participants in the production and transformation of scientific knowledge. Biohacking labs like BioFoundry and Genspace encourage hands-on public engagement with biotechnologies through classes and open workshops, as well as projects concerning local environmental pollution.

Biohackers are also making discoveries that advance our understanding of current scientific problems. From devising coronavirus tests to making science equipment out of everyday items and producing open-source insulin, biohackers are reshaping where scientific innovation can happen.

From Law to Ethics

While biohacking can produce great benefits, the risks can't be neglected. The question is how best to address them.

While laws and regulations are necessary to prevent malicious or dangerous practices, their overuse can also push biohackers underground to tinker in the shadows. Bringing biohackers into the fold of existing institutions is another approach, although this could threaten the ability of biohackers to pose tough questions. In addition to law, ethical codes drawn up by the biohacking community itself offer a productive way forward.

For Stengers, an "ethical" relationship is not based on the domination or capture of one group by another. Rather, it involves symbiotic modes of engagement in which practices flourish together and transform each other.

A balance between law and ethics is necessary. The 2011 code of ethics drawn up by biohackers in North America and Europe is a first step toward what a more open, transparent, and respectful culture of collaboration could look like. In the United States in recent years, there have been experiments with a more open and symbiotic relationship between the Federal Bureau of Investigation and the biohacking community.[5]

But this is just the beginning of a conversation that is in danger of stalling. There is much to lose if it does.

Notes

1. Baumgaertner, E. (2021, October 15). The untold story of how a robot army waged war on COVID-19. *Los Angeles Times*. https://www.latimes.com/world-nation/story/2021-10-15/biohackers-tackle-covid-testing-variants-with-robots.
2. Press, T. A. (2008, April 22). Charge dropped against artist in terror case. *New York Times*.
3. Grushkin, D., Kuiken, T., & Millet, P. (n.d.). *Seven myths and realities about do-it-yourself biology*. Wilson Center. https://www.wilsoncenter.org/publication/seven-myths-and-realities-about-do-it-yourself-biology-0.
4. Stengers, I. (2018). *Another science is possible: A manifesto for slow science* (S. Muecke, Trans.). Polity Press. (Original work published 2013).
5. Wolinsky, H. (2016). The FBI and biohackers: An unusual relationship. *EMBO Reports*, *17*(6), 793–796. https://doi.org/10.15252/embr.201642483.

Contributors

NATHAN AHLGREN is an aquatic microbial ecologist who studies how environmental factors and interactions between microbes and viruses shape the evolution, diversity, and structure of microbial communities. Marine microbial communities in particular are extremely diverse and impact globally important nutrient and carbon cycles. A key aspect to understanding the importance of microbes to our planet is knowing what factors control and maintain the diversity and structure of these communities. Professor Ahlgren uses traditional culture isolation and laboratory studies, along with DNA sequencing and bioinformatics, to elucidate these important abiotic and biotic microbial interactions.

IVAN ANISHCHENKO did his graduate work in computational biology at the University of Kansas in the laboratory of Ilya Vakser, where he worked on developing computational methods for studying protein-protein interactions. He did his postdoctoral work with David Baker of the University of Washington, Seattle, studying coevolution in proteins and building and applying deep-learning methods to the prediction and design of protein structure. Anishchenko's research interests lie in developing computational tools to facilitate biological discovery.

TRINE ANTONSEN is a senior researcher in the Climate and Environment Department at the Norwegian Research Centre (NORCE). Dr. Antonsen holds a PhD in Philosophy from the University of Oslo. Antonsen managed the ReWrite Project (2018–2023), funded by the Research Council of Norway, with the aim of understanding how humans relate to nature and how our perception of nature influences—and is influenced by—emerging biotechnologies, such as genome editing. In NORCE's research group for gene technology, the environment, and society, which is a national competence center for biosafety, Antonsen coordinates research on ethical, legal, and social aspects of biotechnology.

JENNIFER BARFIELD is an Assistant Professor at Colorado State University and has a background in animal science and conservation biology. Her lab currently researches comparative assisted reproductive technologies (ARTs) with a focus on large animals. She leads scientific study for the Laramie Foothills Bison Conservation Herd, a herd for which ART was used to return bison with Yellowstone genetics to open public space in northern Colorado. She created and codirects a one-year, non-thesis master's program in ART, which trains future embryologists and reproductive professionals for careers in human infertility, animal reproduction, and conservation.

PEDRO BELDA-FERRE is an Assistant Project Scientist in the Knight Lab at the University of California, San Diego. Microbes living in our bodies play an important role in human physiology, and several diseases have been linked to microbial dysbiosis. How modern lifestyles influence the human microbiome and health is one of his main interests. He has been involved in projects aimed at developing microbiome-modulating strategies for caries disease and food allergies by using different molecular, bioinformatic, and animal-model methodologies.

ARI BERKOWITZ, PhD, is a Presidential Professor of Biology, the Director of Graduate Studies for Biology, and the Director of the Cellular & Behavioral Neurobiology Graduate Program at the University of Oklahoma. His research focuses on how the spinal cord selects among and generates leg movements. He is the author of *Governing Behavior: How Nerve Cell Dictatorships and Democracies Control Everything We Do*.

ADELINE BOETTCHER completed her PhD in Molecular and Cellular Biology in 2019 at Iowa State University, where her work focused on the characterization and development of biomedical models using SCID (immunodeficient) pigs. She then held a postdoctoral fellowship at Northwestern University, where she studied immunotherapies for prostate cancer in mouse models. She was previously a Technical Writer at Iowa State University and a Scientific Editor for the Radiological Society of North America. Dr. Boettcher launched and ran Alpha Beta

Scientific Communications LLC in 2021 and consulted with biotechnology startups. Boettcher is now an Operations Specialist at Red Cell Partners, an incubation firm, where she works with health care and national security startup companies.

JASON A. DELBORNE is a Professor of Science, Policy, and Society (in the Department of Forestry and Environmental Resources); the Director of the Science, Technology, and Society program at North Carolina State University; and a founding member of the Genetic Engineering and Society Center (also at NC State). He conducts research on public and stakeholder engagement surrounding emergent environmental biotechnologies. He has served on two expert committees for the National Academies of Sciences, Engineering, and Medicine (gene drives and forest biotechnology); a Task Force on Synthetic Biology and Biodiversity Conservation for the International Union for Conservation of Nature; and an expert committee on genetic engineering for the Forest Stewardship Council.

KEVIN DOXZEN received his PhD in Biophysics from the University of California, Berkeley, for his work in the lab of Jennifer Doudna. He is a Hoffmann Fellow jointly appointed in the Thunderbird School of Global Management and the Sandra Day O'Connor College of Law, both of Arizona State University, and the World Economic Forum's Health and Healthcare Platform. Doxzen's research focuses on policy and governance challenges in precision medicine and health care, including equitable access to gene therapies and other breakthrough technologies in low- and middle-income countries.

MO EBRAHIMKHANI is an Associate Professor in the Department of Pathology, School of Medicine, University of Pittsburgh. He is also a member of the Division of Experimental Pathology and the Pittsburgh Liver Research Center. Prior to his current position, he was an Assistant Professor in the School of Biological and Health Systems Engineering at Arizona State University and an adjunct faculty member of medicine at the Mayo Clinic. He performed his postdoctoral training in the Department of Biological Engineering at the Massachusetts Institute of Technology.

ELEANOR FEINGOLD is a Professor in the Department of Human Genetics, Graduate School of Public Health, University of Pittsburgh. Her research is in statistical genetics and genetic epidemiology. She develops statistical methods for analyzing data from new genomic technologies, and she studies the genetics of such varied traits as cleft lip and palate dental health, Alzheimer disease, Down syndrome, and meiotic recombination.

A. CECILE J. W. JANSSENS,† held an MA, MSc, and PhD; her degrees were in economics, psychology, and epidemiology. She was a Research

Professor of Epidemiology at Emory University's Rollins School of Public Health. Her work focused on research methodology, with a specific interest in the (genetic) prediction of common diseases and traits. She was a columnist for the Dutch newspaper *NRC* and the *Dutch Journal of Medicine* (*Nederlands Tijdschrift voor Geneeskunde*). She passed away in September 2022.

SAMIRA KIANI is an Associate Professor of Pathology and Bioengineering at the University of Pittsburgh who works on strategies for safe gene therapy. She has made transformative innovations in the area of CRISPR gene editing. But her work extends beyond the laboratory. In addition to designing genes, she designs experiences that allow people to reflect on the purposeful manipulation of biological systems and to practice a science that is soulful, self-aware, human-centered, and informed by the collective wisdom of society. She describes herself as an integrative designer, one who is an anthropologist at heart, a social scientist in mind, and a biological designer in practice.

AMANDA KOWALCZYK is an evolutionary genomics researcher who studies a broad range of traits relevant to human health, such as adaptations in skin, hair, vision, and life span. She primarily uses computational methods to analyze evolution-based data. She connects basic science research in evolution to clinically relevant outcomes. Kowalczyk earned her PhD in Computational Biology in 2021 through the CPCB PhD Program jointly hosted by the University of Pittsburgh and Carnegie Mellon University.

MARIANA LAMAS is a Research Associate at the Centre for Culinary Innovation at the Northern Alberta Institute of Technology (NAIT). Her work at the Centre focuses on developing new food products and managing and providing planning expertise to projects. She is particularly interested in plant-based food and how new technologies and innovations are changing how we perceive food and the way we eat. Lamas has a Diploma in Culinary Arts from NAIT. She has been working with food research for the past six years. Before that, she worked in the museum and heritage sector and holds a Bachelor's and Master's in Museology.

ANDREW LAPWORTH is a Senior Lecturer in Cultural Geography at the University of New South Wales in Canberra, Australia. He completed his PhD in Human Geography at the University of Bristol, United Kingdom. His main area of research addresses how people and communities make sense of a world being dramatically reshaped by developments in science and technology and what part the arts can play in communicating and potentially transforming that experience. His current research explores the movements known as do-it-yourself biology and biohacking and their role in facilitating public encounters and engagement with biotechnologies.

REBECCA MACKELPRANG received her PhD in Plant Biology in 2017 from the University of California, Berkeley, where she studied plant responses to microbial pathogens. Her enthusiasm for sharing science took her away from the lab bench and into a postdoctoral position in science communication with Dr. Peggy Lemaux, a position focused largely on agricultural biotechnology. In 2019 Dr. Mackelprang was a Mass Media Fellow with the American Association for the Advancement of Science and was placed at Ensia, a nonprofit media outlet, where she wrote about the environment. She then accepted a postdoctoral position in science policy with the Engineering Biology Research Consortium, where she is now the Associate Director for Security Programs and focuses on advancing the responsible development and use of biotechnologies.

KATHLEEN MERRIGAN is the Executive Director of the Swette Center for Sustainable Food Systems, Senior Global Futures Scientist, and the Kelly and Brian Swette Professor of Sustainable Food Systems at Arizona State University with appointments in the School of Sustainability, College of Health Solutions, School of Public Affairs, and Morrison School of Agribusiness. Concurrently, she is a Venture Partner at Astanor, a Belgium-based venture capital firm investing in agtech solutions. From 2009 to 2013, Merrigan served as the US Deputy Secretary and Chief Operating Officer of the US Department of Agriculture. Her prior career includes a variety of academic and agricultural policy positions, including in the US Senate, where she wrote the law establishing national standards for organic food.

SAMAN NAGHIEH is a Biomedical Engineer (Systems Designer and Manufacturing Process Engineer) experienced in design, fabrication, verification, validation, failure modes and effects analysis, corrective and preventive action, and project management. He has a PhD in Biomedical Engineering from the University of Saskatchewan (2020). Saman is specialized in advanced manufacturing and bioprinting and has years of experience in the regeneration of bone, cartilage, and nerves, as well as in total knee arthroplasty implants and cardiac cryoablation catheter development. His research interest is tissue-engineered scaffolds, specifically additive manufactured scaffolds for hard and soft tissues. During his studies, he was selected as a distinguished researcher and was awarded numerous prizes. Naghieh was the President of the Engineering Graduate Community Council (2017–2018) and the President of the Public Service Alliance of Canada Local 40004, Union of Graduate and Postdoctoral Workers, at the University of Saskatchewan (2017–2019).

SEAN NEE wrote his chapter while a Research Professor of Ecosystem Science and Management in the Braithwaite Group at Pennsylvania State University. There, Professor Nee pursues the scientific study of consciousness. He has worked with Robert May, John Maynard Smith, and Sunetra Gupta, among others, as well as Victoria Braithwaite, now deceased. His scientific interests include biodiversity, theoretical ecology, evolutionary biology, and mathematics.

DIMITRI PERRIN is an Associate Professor in the School of Computer Science at the Queensland University of Technology (QUT). He is also a Chief Investigator in the QUT Centre for Data Science, co-leading the Health and Biological Systems Domain. Prior to joining QUT, he worked as a Foreign Postdoctoral Researcher in the Laboratory for Systems Biology (RIKEN, Japan) and as an IRCSET (Irish Research Council for Science, Engineering and Technology) Marie-Curie Research Fellow with the Centre for Scientific Computing & Complex Systems Modelling (Dublin City University, Ireland) and the Department of Information Networking (Osaka University, Japan).

CHRISTOPHER J. PRESTON is a Professor of Environmental Philosophy at the University of Montana in Missoula. He works on the ethics of emerging technologies, wildlife, and gender. His award-winning book *The Synthetic Age: Outdesigning Evolution, Resurrecting Species, and Reengineering Our World* (MIT Press, 2018) has been translated into six languages. A sequel, *Tenacious Beasts: Wildlife Recoveries That Change How We Think about Animals*, was published by MIT Press in 2023. The US National Science Foundation, the Templeton Foundation, and the Kone Foundation (Finland) have all supported his work.

JASON RASGON received his PhD in Entomology at the University of California–Davis, where he studied the population dynamics of *Wolbachia* symbionts in mosquito populations. He then conducted postdoctoral research on transposon population biology at North Carolina State University. He is currently a Professor and the Dorothy Foehr Huck and J. Lloyd Huck Endowed Chair of Disease Epidemiology and Biotechnology in the Department of Entomology, the Huck Institutes of the Life Sciences, and the Center for Infectious Disease Dynamics at Pennsylvania State University. Previously, he was an Assistant, then Associate, Professor at the Johns Hopkins Bloomberg School of Public Health. Dr. Rasgon's research interests are in genetic strategies for controlling vector-borne disease, emergent and invasive pathogens, and the development of molecular biology tools.

PENNY RIGGS is an Associate Professor of Functional Genomics in the Department of Animal Science and the Associate Vice President for

Research at Texas A&M University. Her current research focuses on analyzing gene, RNA, and protein expression, function, and signaling, particularly in skeletal muscle. Other work of hers concerns whole-genome sequencing of bacterial pathogens. She is interested in the application of genomic technologies for ensuring food and nutritional security, and human and animal health.

OLIVER ROGOYSKI, an early-career researcher, undertook a BSc in Biochemistry at the University of Sussex, where he gained valuable research experience through a competitive summer studentship funded by the Genetics Society. This led to his subsequent decision to undertake a PhD studying RNA translation and degradation at Brighton and Sussex Medical School. Following this, Rogoyski is now a Post-doctoral Research Fellow working on RNA–protein interactions at the University of Surrey.

GARY SAMORE served in the US government for more than 20 years, seeking to control nuclear arms and prevent their spread, especially in the Middle East and Asia. In that capacity, he served both President Clinton and President Obama as the senior official in the National Security Council who was responsible for the nonproliferation of weapons of mass destruction. Outside government, he held senior research and administrative positions at the International Institute for Strategic Studies in London, the Council on Foreign Relations, and the Belfer Center for Science and International Security at Harvard University. He holds an MA and PhD from the Department of Govern-ment of Harvard University.

SAHOTRA SARKAR is a Professor in the Departments of Philosophy and of Integrative Biology at the University of Texas at Austin. He obtained his BA from Columbia University and his MA and PhD from the University of Chicago. Professor Sarkar specializes in the history and philosophy of science, environmental philosophy, conservation biology, and disease ecology. He authored the books *Genetics and Reductionism*; *Biodiver-sity and Environmental Philosophy*; *Doubting Darwin? Creationist Designs on Evolution*; *Environmental Philosophy*; and *Systematic Conservation Planning*, along with authoring or coauthoring more than 250 articles, mostly in philosophy and conservation biology. His latest book, *Cut-and-Paste Genetics: A CRISPR Revolution* (Rowman & Littlefield), was published in September 2021.

GEORGE E. SEIDEL,† University Distinguished Professor Emeritus at Colorado State University, was an expert in technologies that enhance animal reproduction in order to propagate valuable genetics. He was a member of the National Academy of Sciences and the National

Academy of Inventors. He was an internationally recognized leader in the biotechnology of preimplantation embryos from cattle and horses. The Seidel Laboratory developed and refined techniques such as nonsurgical recovery and transfer of bovine and equine embryos. Seidel passed away in September 2021.

PATRICIA A. STAPLETON is a Political Scientist at the RAND Corporation. Her research interests include science and technology policy, risk regulation for emerging technologies, risk assessment and communication, and the evaluation of R&D programs. Her past work has examined agricultural biotechnology in the context of food safety and security and the regulation of biotechnologies used in assisted reproductive technology. She holds a PhD in Political Science from the City University of New York Graduate Center and an MA in French Literature from Rutgers University.

CRAIG W. STEVENS is a Professor of Pharmacology at Oklahoma State University–Center for Health Sciences in Tulsa, Oklahoma. He received an MS in Biomedical Sciences from the University of Illinois at Chicago and a PhD in Pharmacology from the Mayo Clinic in Rochester, Minnesota. After a two-year postdoctoral fellowship at the University of Minnesota, he joined the faculty of OSU–CHS in 1990. His research interests include all things opioid; he has published on the topics of the opioid epidemic, the evolution of opioid receptors, and the interaction of the opioid and neuroimmune systems.

PAUL B. THOMPSON is Professor Emeritus of Philosophy at Michigan State University, where he served as the inaugural W. K. Kellogg Chair of Agricultural, Food and Community Ethics until 2022. Thompson's publications on food and agricultural biotechnology have appeared in the *Journal of Animal Science*, *Plant Pathology*, *Nature Biotechnology*, and *Poultry Science*, as well as applied philosophy outlets. From *Field to Fork: Food Ethics for Everyone* was named book of the year for 2015 by the North American Society for Social Philosophy.

CHRISTOPHER TUGGLE holds a PhD in Biochemistry, and his background is in molecular genetics. Currently he is very interested in the genetic and epigenetic control of immunity. He directs a group of fantastic researchers who are studying the structure and function of the pig genome as well as developing a genetic line of immunodeficient pigs as a biomedical model for cancer and stem cell therapies.

VIKRAMADITYA G. YADAV is an Associate Professor at the University of British Columbia, where he directs Canada's premier program in Sustainable Process Engineering. He has made notable contributions to research, education, commercialization, and regulation of synthetic biology and environmental biotechnology. Dr. Yadav also founded Metabolik Technologies Inc., which was acquired by Allonnia, a Bill Gates–backed company, and is currently the Chief Executive Officer of Tersa Earth, a mining biotechnology company. He is also the Chief Technology Officer of React Zero Carbon, a venture catalyst and capital fund for net zero solutions. He was recognized as one of Canada's Top 40 Under 40 in 2021.

MARC ZIMMER is the Jean C. Tempel '65 Professor of Chemistry at Connecticut College and the author of *The State of Science: What the Future Holds and the Scientists Making It Happen* (Prometheus, 2020), *Illuminating Diseases: An Introduction to Green Fluorescent Proteins* (Oxford University Press, 2015), and four books for young adults. His young adult book *Solutions for a Cleaner, Greener Planet* was longlisted for the 2020 AAAS/Subaru SB&F Prize for Excellence in Science Books. His writing has appeared in *USA Today* and the *Los Angeles Times*, and he has been interviewed for and quoted in the *Economist*, *Science*, and *Nature*. Professor Zimmer's research group uses computational methods to study bioluminescent proteins.

Index